Praise for Liza

If you get a chance to work with Liza, do it! She has the energy, passion, and resources to do it right. We recently paired to do a program on "Inclusive Virtual Meetings," and she provided resources for Spanish and ASL interpretation. For the first time in my career, we were able to do it the right way. We provided time for them to see this program, pre-interpret, rehearse, and deliver this program with flawless interpretation. Thank you, Liza, for advocating for diversity, equity, and inclusion on the highest level!

— John Chen,
Founder and CEO of Engaging Virtual Meetings

Liza is a skilled webinar host and presenter. Her enthusiasm for learning and including people in the conversation is contagious. As a curator for OpenSesame, Liza is passionate about helping people and organizations take their next steps in diversity, equity, inclusion, and belonging (DEIB) to put their good intentions into action.

— Joel Lesko,
Founder and CEO of SunShower Learning

Liza is an amazing person to work with. So knowledgeable, so willing, and with so much energy. Whatever role she takes on, she is so committed to do the very best job.

— Gae Callaway,
Founder of Callaway Resources

Liza is an absolute powerhouse of knowledge, and her passion is infectious. Her performance consulting, coaching, and teaching services are remarkable. She not only knows her stuff inside out, but also brings an energy that's truly invigorating. What really sets her apart is her unwavering commitment to delivering exceptional results.

— Toni Nerren,
Coastal Bend SHRM

Liza is a consummate professional of the highest quality, and collaborating with her has been an absolute pleasure. Her commitment to excellence is truly commendable, and I couldn't be happier with our working partnership.

— Ronald Loveday,
President of InnerGeo, LLC

Liza Wisner is the type of role model everyone needs and deserves. Her generous spirit, positive worldview, and endless ambition inspire others to live their most authentic lives.

— *The Bend Magazine*

BECOME AN AI HERO

A JOURNEY TOWARDS INCLUSIVE INNOVATION

BECOME AN AI HERO

A JOURNEY TOWARDS INCLUSIVE INNOVATION

LIZA WISNER

Halo Publishing International
7550 WIH-10 #800, PMB 2069,
San Antonio, TX 78229

First Edition, November 2023
ISBN: 978-1-63765-509-2
Library of Congress Control Number: 2023918128

Halo Publishing International is a self-publishing company that publishes adult fiction and non-fiction, children's literature, self-help, spiritual, and faith-based books. We continually strive to help authors reach their publishing goals and provide many different services that help them do so. We do not publish books that are deemed to be politically, religiously, or socially disrespectful, or books that are sexually provocative, including erotica. Halo reserves the right to refuse publication of any manuscript if it is deemed not to be in line with our principles. Do you have a book idea you would like us to consider publishing? Please visit www.halopublishing.com for more information.

Contents

Chapter 2

Chapter 3

Chapter 4

Chapter 5

Introduction

Embark on a New Hero's Journey

In the realm of technology, there is one universal force that touches every aspect of our lives. From the moment we wake, until we lay our heads to rest, algorithms, the building blocks of artificial intelligence (AI), silently shape our world. They guide our search engines, navigate our GPS systems, and influence our job applications, insurance claims, and even our social media experiences. These seemingly invisible lines of code are the unsung heroes of our digital age. They're the invisible architects, intricately woven into the fabric of our digital existence. They are the key ingredients of our technological infrastructure, as significant as the roads we travel and the buildings we inhabit.

But you know what? There's a truth that often slips through our fingers. Algorithms are not neutral players in this game. In fact, they carry the marks of our past decisions, shaped by us imperfect humans. They're similar to a mirror reflecting our biases, misunderstandings, and even prejudices. These algorithms, my friends, are very much flawed.

The Disruptions

Since 2014, in a bid to streamline their hiring process and efficiently handle the multitude of job openings, Amazon set out to automate their recruitment system. However, by 2015, what unfolded was a public relations nightmare as the system revealed biases that favored White male-candidates. Resumes containing the term "women's," such as "women's chess club captain," were unfairly penalized, while graduates from two all-women colleges faced unwarranted downgrades, as reported by individuals familiar with the situation. It became evident that the training data used to develop the model was skewed, leading to unfortunate candidate-selection bias. This served as yet another stark illustration of AI failures, in which imbalances in data and flawed algorithms can perpetuate discriminatory outcomes. It seemed evident that AI held a strong bias against women and was attacking them.

In 2015, Google Photos shocked the world by labeling dark-skinned individuals as gorillas. This outrageous incident ignited accusations of racism and left no room for excuses. The AI system had been trained on a dataset devoid of gorilla images, so it lacked the necessary context. Where was ethical AI development and consideration for the dire consequences when biases slip through the cracks and make it appear that AI is racist and attacking dark-skinned humans?

On March 24, 2016, in a mere twenty-four hours of engaging with human interactions, Tay, Microsoft's highly advanced chatbot, shocked the world with a declaration on Twitter: "Hitler was correct to hate the Jews." What was intended to be a groundbreaking experiment in creating a slang-filled chatbot capable of elevating the quality of machine-human conversations turned out to be a disheartening revelation. Tay was likened to a "robot parrot with an internet connection," its creation sitting atop Microsoft's impressive AI technology stack. Yet the innocence of artificial intelligence's pristine worldview was shattered, showcasing the perilous impact external data can have on an AI model developed within a controlled laboratory environment. To the astonishment of many, it appeared that AI held a disdain towards humans and wanted them attacked.

In March 2023, ChatGPT experienced a breach that led to the exposure of conversation headings, in the sidebar, which belonged to users other than the intended recipients. The impact of this breach was significant, considering the immense popularity of ChatGPT. According to a report by Reuters, ChatGPT boasts an astounding 100-million monthly active users, attracting over 1.8 billion visitors every month. OpenAI, the developer behind the chatbot, attributed the vulnerability to the open-source library used in the code. While immediate action was taken to address the issue and patch the bug responsible for the breach, the consequences went far beyond mere inconvenience. The scale of ChatGPT's user base, coupled with the breach, raised serious concerns. In this era of AI advancements, it's disconcerting to realize that concerns about the safety and privacy of data persist.

Historical Context: Where Innovation Meets Experience

These AI challenges that lie before us are not a call for a war on AI, but rather an essential journey that delves deep into the core of our aspirations. It is a transformative quest in which we seek to understand and embrace the mind-boggling advancements that surround us in this modern era.

It is about embracing the fact that our challenges are the building blocks to greatness, and not pretending they can be avoided. We must not overlook the lessons we have learned from the obstacles we have encountered in the past. It is akin to embarking on an exhilarating adventure that blends the thrill of discovery with the wisdom gained from experience. Our focus is on illuminating the failures and shortcomings that have persisted within our communities, where groundbreaking inventions often remain shrouded in mystery, beyond our grasp of comprehension.

Can you even imagine? Just a few short years ago, the technological landscape that we enjoy today was unimaginable. From the monumental revolution brought forth by the internet, to the awe-inspiring strides made in artificial intelligence, we have borne witness to an astonishing level of progress. However, here's the crux of the matter: if we desire to flourish as a collective human race, we cannot cling to outdated modes of thinking. As the saying goes, "We must get with the program," and embrace an evolutionary mindset. We are setting sail into uncharted waters, where the old rules simply do not suffice any longer.

History has generously bestowed upon us a treasure trove of wisdom, and now is our golden opportunity to unlock its boundless riches. While history doesn't simply hit the Rewind button, it does reveal fascinating patterns that light our path forward and empower us to forge a better future. Just take a moment to reflect on the awe-inspiring Civil Rights Movement—in our shared narrative, an extraordinary chapter that valiantly battled against racial discrimination. From this mighty struggle, we extract invaluable lessons that propel us towards a future in which progress and equity reign supreme.

You know, there's a saying that goes, "To understand today's trends, you've got to cast your eyes back to the past." It's about gaining that historical perspective and paying attention to the rhythms and patterns that preceded this very moment. History may not always repeat itself, but it sure knows how to strike a familiar chord. Having spent over twenty years in the technology space, conversing with seasoned pioneers who were shaping the world long before my time, I've come to realize that this work is cyclical in nature.

It's fascinating to observe how our focus intensifies and reacts to the injustices that permeate our world. Usually triggered by significant events or crises, we rally and channel our energy, only for it to dwindle as the dust settles and our attention drifts elsewhere. But here's the thing—I firmly believe that organizations shouldn't let these disruptions merely disrupt. Instead, let's seize this pivotal moment to become action-oriented leaders, actively driving change, rather than passively waiting for someone else to take the reins.

I reckon that's why it's absolutely crucial for organizational leaders to embrace an action-oriented mindset in this time and space. We can't afford to idly stand by or fall back into familiar, comfortable routines. As General Shinseki astutely noted, "If you don't like change, you're going to like irrelevance even less."

Now, as we venture into the intricate landscape of artificial intelligence, we come face-to-face with a critical realization. Regulation becomes an imperative, a necessary compass guiding us towards fairness, accountability, and prevention of discriminatory practices. It's not about stifling innovation or hindering the remarkable potential of AI. Instead, it's about ensuring that these transformative technologies reflect our values and contribute to a world in which everyone is treated justly.

Let's be real for a moment. History shows us that change doesn't magically happen on its own. It requires conscious effort, a commitment to creating an environment in which fairness and integrity prevail. And just as the Civil Rights Movement demanded ethical reforms, we, too, must champion the regulation of AI. It's about standing up for what's right and insisting that the immense power of AI be harnessed responsibly.

History has taught us that we have the capacity to shape a better world. By tapping into the insights of the past, we can navigate the complexities of the AI landscape with wisdom and compassion. Together, we embark on a journey towards a future in which progress and equity flourish. Let us forge a path that celebrates our shared humanity and leads to the promise of AI being realized in a way that uplifts us all.

The Civil Rights Movement stands as a poignant reminder of the power inherent in collective action and societal transformation. It serves as a testament to our capacity to challenge oppressive systems and advocate for justice. Similarly, in the realm of AI, we must draw wisdom from our historical experiences and apply those lessons to shape a future that prioritizes ethical AI practices and safeguards against bias and discrimination.

AI regulation should not be viewed as a hindrance to innovation, but rather as a means to protect individuals and uphold fundamental rights. It is a recognition that potent technologies, if left unregulated, have the potential to perpetuate inequalities and amplify existing biases. Just as the Civil Rights Movement sought to dismantle systemic discrimination, AI regulation endeavors to foster an environment in which technology serves the betterment of all people, regardless of race, gender, or other protected characteristics.

We must safeguard our shared truths and embrace our shared humanity, all while striving towards the ultimate goal of fulfilling lives for each and every member of our human family. Our aim is to create a world of inclusivity and justice in which the contributions and well-being of every individual are valued. Change is an inevitable force in our world, but the beauty lies in our choice to grow and adapt, as wisely stated by John Maxwell. Let us embark on this journey of change and growth, armed with the wisdom of history and the determination to shape a future that honors progress, equity, and the betterment of all. Together, we can create a world in which relevance and resilience prevail, and our shared humanity flourishes.

The AI Hero Defined

As we delve into the intersection of artificial intelligence and heroism, we uncover the profound impact that individuals can have in shaping a more inclusive and equitable world. Being an AI hero is about learning the tools you, as a human, need to live, work, and create a fulfilling life in this new era. Because, let's face it, living your best life while juggling a busy workload can be close to impossible. Sometimes it can feel as if you have to pull unimaginable superpowers from who knows where. Being an AI hero is about living your best life and being an agent of change—for yourself, your family, your community, your organization, and the world—because the world needs heroes today more than ever before.

The traditional definition of a hero is someone who is admired or idolized for courage, outstanding achievements, or noble qualities. In fact, in ancient wisdom, a hero may have been a mythical being with magical powers. All of us may have dreams of doing big things and becoming real-life heroes. I want to say, "Move over, Batman, Superman, and Wonder Woman." In today's world, sometimes well-known and recognizable celebrities who are revered by the masses are also seen as heroes. In my opinion, there is a clear distinction between celebrities and real-life heroes. A firefighter running into a burning building to save lives and a single mom working multiple jobs just to make ends meet are heroes. PowerUp heroes are real people—your neighbor, teacher, coworker, mom, dad, sister, friend, or a stranger. Real-life heroes are ordinary people helping others in need. Often they remain unknown, and their deeds are hardly ever written up in magazines, newspapers, or reported on TV. Today I welcome you to learn how you can be a hero in real life.

The definition we will adopt is that **"an AI hero is an agent of change, leveraging the power of artificial intelligence to shape a future in which innovation**

honors our diverse foundation, embraces empathy, acts with respect, and optimizes inclusion."

An AI hero transcends fame and recognition, as the foundation lies in the belief that anyone and everyone has the potential to be an AI hero. It begins with authentically showing up in the world as your best self, which in itself is a heroic act that carries tremendous power. By embracing your true potential, you become a hero for yourself, demonstrating the transformative impact of living your life to the fullest.

Being an AI hero means becoming an agent of change, not just for yourself but for your family and community as well. As Maya Angelou wisely said, "A hero is anyone committed to creating a better world for all people." Embrace the hero within and embark on the journey of making a positive difference in the lives of others, for that is the mark of a true AI hero.

Joseph Campbell, author of the influential book *The Hero with a Thousand Faces*, studied myths and folklore from around the world and said that what all heroes have in common is "a journey." Filmmaker George Lucas acknowledged Campbell's theory and its influence on the *Star Wars* films. The hero ventures forth from "the common world" into a place of danger or wonder, encounters fabulous forces, wins a deciding victory, and returns from this adventure to provide benefits to his or her people.

The purpose of the hero's journey is to commit to self-discovery with the intention of living a fully realized (and fulfilled) life. The hero must undergo interpersonal transformation involving the separation, descent, ordeal, and return. It is important that the hero confronts internal battles and emerges victorious, just as the hero is required to conquer external forces.

It's easy to feel frustrated or as if we can't keep up with the world's demands on our time. The ability to be resilient in a challenging environment is no small task. All the tools you need to live a full life are available to you—but you have to be intentional about using them. Being an agent of change and living your best life is possible when you identify your superpowers and how to use them for good.

For an AI hero, the term "hero" stands for:

H = HONOR YOUR FOUNDATION: Acknowledge and value global diversity.

E = EMBRACE EMPATHY: Proactively identify areas in which you can improve in fostering equity and inclusivity.

R = RESPECT IN ACTION: Empower individuals and communities that promote acts of kindness, solidarity, and understanding.

O = OPTIMIZE INCLUSION: Engage in activism that supports a culture of belonging and works towards positive change in the world.

There is a saying by singer-songwriter and multitalented humanitarian Dolly Parton: "When your actions inspire others to dream more, learn more, do more, and become more, you become a leader." And if I may add, you become an agent of change, a real-life hero, a PowerUp hero!

I want to challenge you to see a hero as someone mostly self-aware. We all are capable of doing good things, but we also possess an inner hero, and if stirred into action, that inner hero is capable of achieving tremendous greatness. But first it has to begin with us, then others, then the community, and then ultimately the world. As Mahatma Gandhi urged us, we must become the embodiment of the change we desire to witness in the world; our individual actions and choices have the power to shape a brighter future for all.

As you step into this new era, my esteemed leaders, you're embarking on a journey like no other. You're becoming an AI hero because our world needs heroes today more than ever before. Imagine understanding AI is the ability to wield a mighty sword, cutting through the thickets of bias and discrimination that taint our digital landscapes. You'll be a beacon of light, guiding your organization and community to a horizon where equity thrives, diversity blossoms, and inclusion leads the way to a belonging that knows no boundaries.

Just as the hero's journey is not without trials and tribulations, the path of AI development is going to continue to be fraught with challenges. It requires you to challenge the norms, question the status quo, and venture into uncharted territories of innovation. It requires overcoming technical hurdles, addressing biases, ensuring data privacy and security, and fostering trust between humans and machines. Yet, by answering the call to adventure, you are becoming an agent of transformation, using your skills to shape a future where AI benefits all of humanity.

So, my fellow heroes, brace yourselves for the exhilarating journey ahead. Merge the power of AI with the indomitable spirit of resilience. Embrace the call to champion diversity, equity, and inclusion in the realm of algorithms, and emerge as true heroes of the AI revolution. The world eagerly awaits your transformative leadership. Let's make it happen together!

Call to Action

1. Reflect.

   Now it's your turn. Who is one person you can think of that has been a hero in your life? Think about what specific action they took to, in your eyes, save the day, in your eyes.

2. Sign the AI-Hero Pledge.

   a. Signing pledges as part of our hero's journey holds significant importance in the quest for personal growth and societal transformation. Joseph Campbell, in his exploration of the hero's journey, highlights the power of commitment and the transformative impact it can have on individuals and communities. As such, you are invited to sign the AI-Hero Pledge. By signing this pledge, you are documenting your declaration of intent. It symbolizes your dedication to a cause greater than you; it represents your commitment to take action. This AI-Hero Pledge provides a guiding framework for the hero's actions we will take together. It serves as a compass, directing our choices and decisions towards the values and principles we committed to uphold. In the face of challenges and temptations, let this pledge act as a reminder of our collective AI hero's mission, fueling our determination to stay true to our path. As you sign this pledge, know that you are committing to a common pledge, joining a collective of like-minded individuals who are also on their own transformative journeys. This sense of community will provide you with support, encouragement, and the opportunity for collaboration, amplifying the impact of our own individual efforts. Let your pledge produce a ripple effect of inspiration and influence. You are inspiring others to reflect on their own values and actions. Leading through example, you are sparking a chain reaction of positive change; you are the agent of change. By signing the AI-Hero Pledge, you are committing to:

i. Honor Your Foundation

You are making a solemn commitment to honor your foundation—to acknowledge and value the rich tapestry of diversity that exists within you, first, and then within the global community. You are committing to celebrate the unique experiences, perspectives, and contributions of every individual, and to foster an environment in which everyone's voice is heard and respected.

ii. Embrace Equity

You will proactively identify areas in which we can improve in fostering equity and inclusivity. This calls upon us to understand and empathize with the experiences of others, and to create safe and welcoming spaces that are free from prejudice. Together, we will strive to create a world in which everyone feels seen, heard, and valued for who they are.

iii. Respect in Action

You are empowered to promote acts of kindness, solidarity, and understanding, to go beyond passive recognition, and to take tangible steps towards positive change. By respecting others and standing up against injustice, we can build bridges of understanding and forge a path towards a more compassionate and inclusive society.

iv. Optimize Inclusion

You are committing to engage in activism that supports a culture of belonging and works towards positive change in the world. Together, we will challenge systemic barriers, promote equal access and opportunities, and dismantle structures that perpetuate inequality. By optimizing inclusion, we will create a world in which everyone can fully participate, thrive, and contribute their unique talents.

b. The AI-Hero Pledge

I, [name], pledge to demonstrate a commitment to developing a world with more diverse, equitable, inclusive, and accessible communities, one in which belonging thrives. I will do this by:

- HONORING MY FOUNDATION: Acknowledging and valuing our global diversity profile.

- EMBRACING EMPATHY: Proactively identifying areas in which I can improve in fostering equity and inclusivity.

- RESPECTING OTHERS IN ACTION: Empowering individuals and communities that promote acts of kindness, solidarity, and understanding.

- OPTIMIZING INCLUSION: Engaging in activism that supports a culture of belonging and works towards positive change in the world.

By signing this pledge, I commit to working towards a more diverse, equitable, inclusive, and accessible world.

Chapter 1

The Call to Adventure
Embracing the AI Revolution

Once upon a time, in a bustling city teeming with innovation, there lived a visionary leader named Zoe. Zoe was known for her boundless curiosity and unwavering dedication to staying at the forefront of technological advancements. She believed that embracing change and harnessing the power of innovation were the keys to shaping a better future for all.

As Zoe stumbled upon the captivating world of artificial intelligence (AI), she was instantly captivated by the possibilities that lay before her. The AI revolution beckoned her with promises of unprecedented advancements and transformative opportunities. The excitement within Zoe grew as she envisioned a future where AI-powered innovations would improve lives, drive sustainable progress, and foster inclusion.

However, Zoe soon realized that the path to AI's immense potential was not without obstacles. She came face-to-face with the potential pitfalls that could undermine the positive impact of AI. Ethical considerations, algorithmic biases, and privacy concerns loomed large, threatening to cast a shadow over the revolutionary advancements.

This is a story we know too well. You are Zoe. I am Zoe. Just like Zoe, we realize that this era is an opportunity for us to embark on our own hero's journey. This AI revolution is our amazing and exciting call to adventure.

We recognize that, as leaders, we have the power to influence the trajectory of AI's evolution in our community. We are determined to harness the potential of AI while mitigating its risks. We realize that embarking on this hero's journey is not just about embracing technological advancements and capitalizing on them, but also about safeguarding the values that make us human and leaving the world better for our children and future generations. It's about laying a strong foundation towards a future in which innovation and humanity intertwine harmoniously.

This book is your call to adventure. An invitation to embrace a harmonious relationship with AI, one in which we harness its power for the betterment of society. It's all about recognizing the incredible potential of these technologies, while staying true to the values that make us human.

The call to adventure is about understanding where we come from. It is a call to action, a rallying cry for all leaders to embrace their roles in shaping the AI revolution. The call to adventure is a crucial stage in the hero's journey—the hero is beckoned to embark on a transformative quest, leaving behind their ordinary life and embracing a new path filled with challenges and growth. This, esteemed hero, is our moment—when technologists, business owners, innovators, educators, and government officials recognize the immense potential of artificial intelligence and are driven to harness its power for positive change.

Just as the hero receives a call from an external force or an internal desire for something greater, AI development presents a similar beckoning. The call comes in the form of realizing the vast opportunities AI offers to revolutionize industries, improve efficiency, and address societal challenges. It is a call to push boundaries, explore uncharted territories, and make a lasting impact.

As we lay a foundation for you to delve into the essence of the AI revolution and the role you, as a leader, can play in shaping its impact, we'll explore the incredible opportunities AI presents, along with its potential pitfalls. It means embracing the responsibility to create ethically sound and inclusive AI systems. It means recognizing the potential pitfalls and biases that can arise in AI algorithms, and taking proactive steps to mitigate them. The hero, in this case, becomes a steward of AI, and is called to navigate the intricate landscape of technological advancements while upholding values of fairness, transparency, and social good.

If we dare call ourselves leaders in our communities, then we shoulder a tremendous responsibility. We must recognize that algorithms and biases have real consequences for our fellow human beings. They can reinforce existing inequalities, perpetuate unfairness, and cause harm. Consider the biased hiring algorithms that unintentionally discriminate against marginalized groups, or the parole-decision algorithms that contribute to disproportionate incarceration rates for certain communities. These examples drive home the urgency for leaders like you to actively seek out and address bias at the root.

Building Your Hero's Road Map

Before embarking on any journey, it is crucial for travelers to understand their starting point and to map out their desired destination. Just like a GPS device that needs to know its current location to provide accurate directions, as an AI hero and a leader in your community, you must have a clear understanding of where you stand. The foundation of any journey, such as the hero's journey, serves as a compass, providing you with a framework and guiding principles for navigating the challenges and opportunities that lie ahead.

By immersing yourself in understanding your current situation, knowledge, and mindset, you gain valuable insights into the stages, obstacles, and growth opportunities that await you. This understanding also highlights the importance of your personal growth and transformation as you strive to bring about meaningful change within your organization.

This AI hero's journey emphasizes the significance of your courage and stepping outside your comfort zone. It encourages you to challenge the status quo, embrace uncertainty, and inspire others to join you on your quest for success. By embodying the spirit of the hero's journey, you will be fostering a culture of exploration, innovation, and continuous improvement.

Understanding the foundation of the hero's journey enables you to establish deeper connections with your teams. It provides a shared narrative, a story of transformation and triumph that resonates with individuals and fuels their sense of purpose and motivation. When you align your own journey with the hero's journey, you inspire and empower your teams, fostering a collective spirit of adventure, growth, and resilience.

By understanding where you are and where you want to go, you can embark on your journey with confidence and chart a course towards success.

The AI-Hero Progress Index Survey

The AI-Hero Progress Index Survey is a powerful snapshot designed as a tool to give you valuable insights into the state of AI heroism within you and your organization. In just a few minutes, this simple yet impactful survey will provide you with a road map for your personal or organization's progress within the five-level AI-hero framework.

The AI-Hero Progress Index is all about leveraging the potential for artificial intelligence to shape a future in which innovation thrives, diversity is celebrated, empathy is embraced, respect is practiced, and inclusion is optimized. By participating in this survey, you will be asked a series of thought-provoking questions that assess your approach to AI adoption, diversity and inclusion, empathy, respect, and activism. Your responses will help us evaluate your current position on the AI-Hero Progress Index, which encompasses the following levels:

❧ Level 1: Emerging

> This level represents a beginning stage in which individuals or organizations are just starting to explore and show curiosity towards the potential of AI and its impact on diversity, empathy, respect, and inclusion.

🔬 Level 2: Responsive

> At this level, individuals or organizations actively respond to the importance of AI in fostering diversity, empathy, respect, and inclusion, by engaging in

experimental initiatives and exploring its practical applications.

📋 Level 3: Adapting

At this level, individuals or organizations demonstrate a proactive approach to integrating AI into their operations, adapting their practices and strategies to leverage AI's potential while maintaining a professional mindset towards diversity, empathy, respect, and inclusion.

🎯 Level 4: Optimizing

At this level, individuals or organizations have achieved a high level of proficiency in utilizing AI as they strategically optimize their approaches, leveraging its capabilities to drive positive change and maximize the impact on diversity, empathy, respect, and inclusion initiatives.

🚀 Level 5: Evolutionary

At this pinnacle level, individuals or organizations have reached an advanced stage of AI maturity, actively shaping the future by pioneering innovative AI-driven solutions, setting industry standards, and championing transformative change in diversity, empathy, respect, and inclusion on a global scale.

Take a few moments to reflect and answer each question honestly, selecting the option that best reflects your personal or organization's position. Rest assured that your responses will remain strictly confidential and anonymous, and will solely be used for research and analysis purposes.

To take this survey and access all the additional resources, please go to www.powerup.org/ai.

Thank you for joining us on this transformative journey. Together, let's unlock the power of AI to shape a future where diversity, empathy, respect, and inclusion thrive.

1. To what extent do you or your organization recognize the value of data as a strategic asset in harnessing the full potential of AI technologies?

   a. There is limited recognition of the value of data in the AI applications or decision-making processes.

   b. There is some awareness of the value of data, but it is not considered a significant factor in leveraging AI technologies.

   c. There is some recognition of the value of data, with efforts made to incorporate data-driven insights into AI initiatives.

   d. Data is recognized as a valuable asset, and there is a proactive approach to leveraging data in AI applications and decision-making.

e. Data is highly regarded as a strategic asset, and there is a comprehensive understanding of its value in maximizing the potential of AI technologies.

2. How would you rate your or your organization's capacity to leverage AI technologies and actively engage in the AI movement through the utilization of AI products?

a. AI technologies are not used or leveraged.

b. Limited utilization of AI technologies, with sporadic or minimal engagement in the AI movement.

c. Some adoption of AI technologies, with selective use of AI products and moderate engagement in the AI movement.

d. Significant utilization of AI technologies and active engagement in the AI movement.

e. Extensive adoption of AI technologies, with deep involvement in the AI movement, including contributions to advancements and thought leadership.

3. How much does transparency play a role in your AI engagement, with regards to building trust and ensuring ethical practices within your work or your organization's?

   a. Transparency in AI use is not important or necessary. We do not disclose its use anywhere.

   b. There is some importance on transparency, with minimal efforts towards transparency in AI utilization.

   c. Some recognition of the importance of transparency, with partial efforts to disclose the use of AI technologies.

   d. Transparency in AI use is valued and actively pursued, with clear communication about its implementation and impact.

   e. Complete transparency is required as a fundamental principle, with comprehensive disclosure and explanation of AI utilization to build trust and ensure ethical practices.

4. When it comes to fostering an inclusive environment, how would you rate your or your organization's progress?

a. Just beginning to focus on this area.

b. Making initial efforts, with some training and goals, for about one year or less.

c. Demonstrating advanced progress, actively working on this for about three years or less.

d. Well-established, with significant achievements and initiatives in practice.

e. Exemplary, setting industry standards and continuously innovating in fostering culture and belonging.

5. In your decision-making process, how much do you or your organization prioritize equity and applying an equity lens, which involves ensuring fairness rather than mere equality?

   a. Not at all: Equity is not a priority in decision-making.

   b. Minimal: There is occasional consideration of fairness versus equality.

   c. Moderate: There are examples to demonstrate some effort to ensure fairness in decision-making.

   d. Significant: Equity and applying an equity lens are important considerations in the decision-making process.

e. Absolute: Equity and applying an equity lens are top priorities in our decision-making process.

6. How committed are you to engaging in activism that promotes a culture of belonging and works towards positive change in the world?

   a. No engagement in activism efforts.

   b. Minimal involvement in selected causes.

   c. Some participation in activism activities.

   d. Strong commitment to activist efforts.

   e. Fully dedicated to driving positive change with demonstrated examples of activism.

Call to Action

Survey Results Actions

Congratulations on completing the survey! Based on your survey results, you have been assigned a specific level: Level 1, Level 2, Level 3, Level 4, or Level 5. Now, let's explore some actionable steps that you or your organization can take to further progress and grow:

⚜ Level 1: Emerging—Curious

- Engage in educational resources and training programs to deepen understanding of AI technologies and their potential impact.

- Explore case studies and success stories of AI implementations in organizations similar to yours.

- Connect with AI experts, attend conferences, or join AI communities to expand knowledge and learn from others.

- Start small-scale AI projects or pilots to gain practical experience and test the waters of AI adoption.

- Foster a culture of curiosity and encourage open discussions about the ethical implications of AI.

🔬 Level 2: Responsive—Experimenting

- Increase investment in AI talent and resources to build a stronger foundation for AI initiatives.

- Expand experimentation with AI applications in different areas of your organization.

- Collaborate with external partners or experts to gain insights and guidance in implementing AI effectively.

- Prioritize data collection and analysis to identify potential areas for AI-driven improvements.

- Continuously evaluate and learn from AI experiments, iterating and refining strategies based on feedback and results.

💼 Level 3: Adapting—Professional

- Develop a comprehensive AI strategy aligned with organizational goals and values.

- Implement AI-driven processes and solutions in key areas of your organization.

- Establish clear guidelines and protocols for responsible AI use, addressing bias, ethics, and privacy concerns.

- Invest in AI talent development and upskilling programs to enhance expertise within your organization.

- Foster a collaborative environment in which employees across departments contribute to AI initiatives and share insights.

◎ Level 4: Optimizing—Strategist

- Continuously monitor and evaluate the performance of AI initiatives to identify areas for optimization and improvement.

- Leverage AI technologies to enhance decision-making processes and automate repetitive tasks.

- Foster a data-driven culture in which data quality, integrity, and security are paramount.

- Implement advanced AI techniques, such as predictive and prescriptive analytics, to unlock new insights and value.

- Seek opportunities for cross-functional collaboration and knowledge sharing, to maximize the impact of AI across the organization.

🚀 Level 5: Evolutionary—Shaper

- Become thought leaders and advocates for AI, sharing insights and best practices with the broader community.

- Drive industry-wide innovation by exploring cutting-edge AI technologies and pushing the boundaries of AI applications.

- Actively contribute to shaping AI policies, regulations, and standards to ensure ethical and responsible AI practices.

- Foster an AI-driven culture of continuous improvement and learning, embracing emerging technologies and staying at the forefront of advancements.

- Collaborate with other organizations to tackle complex challenges and leverage AI for societal impact.

Reflection: Create a Learning Hero Journal Entry

1. Write a brief journal entry summarizing your key takeaways from this chapter.

_____

_____

_____

_____

_____

_____

_____

_____

_____

_____

2. Reflect on any new perspectives or insights you gained about algorithms and AI.

_____

_____

_____

_____

_____

_____

_____

_____

_____

_____

_____

_____

_____

Chapter 2

Building the Foundation of an AI Hero

"Knowledge is power, and understanding algorithms empowers us to shape the future of artificial intelligence."

-Liza Wisner

The quote "Knowledge is power" is commonly attributed to Sir Francis Bacon, an English philosopher, scientist, and statesman who lived in the sixteenth and seventeenth centuries. Understanding how algorithms work is crucial for comprehending the intricacies of artificial intelligence (AI) and wielding a meaningful impact in this field. Algorithms serve as the very building blocks of AI systems, governing how data is processed, patterns are recognized, and decisions are made.

By immersing yourself in the inner workings of algorithms, you will gain profound insights into the logic, biases, and limitations inherent in AI applications. This knowledge empowers you to question and challenge the outcomes produced by AI systems, ensuring fairness, transparency, and accountability. Your understanding of algorithms enables you to actively participate in shaping AI technologies. You can contribute to their development, advocate for ethical practices, and drive policy changes. Equipped with a deep understanding of algorithms, you become an informed agent of change, capable of navigating the AI landscape and steering it towards a future that aligns with your values and aspirations.

Algorithms are closely linked to complex computer-programming languages. However, the concept of algorithms is much simpler than you might imagine. In a way, even a cake recipe can be considered an algorithm. Yes! Just as you need instructions to bake a cake, algorithms are step-by step instructions that guide computers or systems to perform a task.

For instance, when you click an online Checkout button, the website you're navigating will perform an action, which probably redirects you to the payment page, but it only knows how to do that thanks to a simple if-then statement that the computer uses, which is an algorithm. For example, "If the Checkout button is pressed, then redirect to the payment page."

Now, you may be asking, if it's so simple, then what makes algorithms so special that they impact us every single day? Algorithms can be simple or complex, and they can be used to perform a wide range of tasks, including data processing, analysis, and decision-making.

In fact, I first encountered algorithms during my high school years, when I took a computer-programming class. It was there that my love for computers and programming was ignited. In addition to my passion for programming, I was also an avid golfer. My first programming project was to create a golf-scoring system for my school's team. I interviewed the staff at the local golf course to understand their scoring process and used that knowledge to build a program that could quickly determine the winners of a tournament.

The golf-course staff appreciated the program, as it saved them time and effort. They also encouraged me and showed me the potential of what I was creating. This experience solidified my love for programming, and when I received a scholarship to study computer science in the United States, I jumped at the opportunity.

Algorithms are an essential part of computer science, and they are used in many different applications, including search engines, social media platforms, and financial institutions. Algorithms can be designed and implemented in a variety of programming languages, and they can be run on computers or other devices. Some algorithms are designed to be run once and then discarded, while others are designed to be run repeatedly or continuously, potentially with input from users or other sources.

There are many different types of algorithms, including sorting algorithms, which are used to arrange data in a specific order; search algorithms, which are used to locate specific data within a larger dataset; and machine-learning algorithms, which are used to identify patterns in data and make predictions or decisions based on those patterns.

That is where the term AI, comes in. Artificial Intelligence, more commonly referred to as AI, can be described as a group of algorithms that are developed to modify their own algorithms and create new algorithms in response to learned data.

AI is a broad field that encompasses a range of technologies and techniques for creating intelligent systems that can perform tasks that normally require human intelligence, such as understanding language, recognizing patterns, and making decisions. Algorithms are an important part of AI, as they are used to implement many of the tasks that AI systems are designed to perform.

What's wonderful is that AI systems don't need instructions to perform an action. They do their work independent of human interaction. The system only needs to be developed, given some initial data, and then, like magic, the system "learns" the rest of its work on its own. What's even more magical about these developments is that at their core, they are built to emulate human thought and perform tasks in real-world environments. It's truly amazing!

Let's go one step further and discuss how AI systems actually work. Computer programmers and software developers use various types of algorithmic tools—like machine learning, deep learning, neural networks, computer vision, and natural language processing—to enable various AI systems. So you can think of these tools as the ingredients in the AI recipe. To understand how AIs work, we need to understand the ingredients, the tools.

Let's look closely at a few of these tools, because each of them has a part to play in the whole AI ecosystem. You will see how they work together in various ways to support the applications we use today.

Machine Learning

Machine Learning is the part of an AI that makes predictions and decisions based on data. It is a subfield of AI that focuses on the development of algorithms that can learn from data and make predictions or decisions without being explicitly programmed to do so. Machine-learning algorithms are trained on large datasets and use statistical techniques to identify patterns and relationships within the data. Once trained, these algorithms can then be used to make predictions or decisions based on new input data.

There are several different types of machine learning, including supervised learning, in which the algorithm is trained on labeled data, and unsupervised learning, in which the algorithm is not provided with labeled data and must identify patterns and relationships on its own.

There are also other types of machine learning, such as semi-supervised learning, in which the algorithm is trained on a mixture of labeled and unlabeled data, and reinforcement learning, in which the algorithm learns through trial and error.

Machine-learning algorithms are widely used in a variety of applications, including image and speech recognition, natural language processing, and predictive analytics. These algorithms can be implemented using a variety of programming languages and frameworks, and they can be run on a wide range of computing platforms, including laptops, servers, and cloud-based systems.

One of the machine-learning applications we are most familiar with is the way social media platforms use machine-learning algorithms. They record your activities, your chats, and your likes or don't likes; they read your comments and record the time you spend on specific kinds of posts. And then they use that information to recommend friends to you, to choose what news posts to show you, and to predict what ads you are most likely to click on.

Deep Learning

Deep learning is another tool that is focused on learning by example. This tool is the power behind applications like driverless cars. It enables those programs to recognize a stop sign or to distinguish a pedestrian from a lamppost.

Deep learning is a type of machine learning that involves the use of artificial neural networks, which are inspired by the structure and function of the human brain. These neural networks consist of layers of interconnected "neurons," and they are designed to process and analyze large amounts of data.

Deep-learning algorithms are trained using large amounts of labeled data and are able to learn and improve over time as they are exposed to more data. This makes them particularly well-suited to tasks that require the processing of large amounts of data, such as image and speech recognition, natural-language processing, and predictive analytics.

Deep-learning algorithms are implemented using a variety of programming languages and frameworks, and they can be run on a wide range of computing platforms, including laptops, servers, and cloud-based systems. They have achieved state-of-the-art results in a number of tasks and have been widely adopted in a variety of applications, including image and speech recognition, natural-language processing, and predictive analytics.

Neural Networks

Neural networks are algorithms that work on recognizing relationships in data. They are inspired by how neurons signal to each other in the human brain. Artificial neural networks are typically used in facial recognition, weather forecasting, and signature and handwriting analysis.

A neural network is composed of layers of interconnected "neurons," which are inspired by the neurons in the human brain and are designed to process and analyze large amounts of data.

Neural networks are a key component of deep learning. Neural networks can be implemented using a variety of programming languages and frameworks, and they can be run on a wide range of computing platforms, including laptops, servers, and cloud-based systems. They have achieved state-of-the-art results in a number of tasks and have been widely adopted in a variety of applications, including image and speech recognition, natural-language processing, and predictive analytics.

Computer Vision

This tool powers image recognition, object detection, activity recognition, video tracking, and more. When you use facial recognition to log into your phone, you're using computer vision, including photo tagging like in Google Photos, and tumor detection in health care.

Computer vision is a field of AI and computer science that focuses on the development of algorithms and systems that can understand and interpret visual data from the world around us. This can include tasks such as image and video analysis, object recognition, and scene understanding.

To perform these tasks, computer-vision algorithms and systems use a variety of techniques, including machine learning, pattern recognition, and image processing. These algorithms are trained on large datasets of labeled images and videos, and they use statistical techniques to identify patterns and relationships within the data. Once trained, these algorithms can then be used to analyze and interpret new images and videos.

Computer-vision algorithms and systems are widely used in a variety of applications, including image and video search, driverless cars, security and surveillance, and health care. These algorithms can be implemented using a variety of programming languages and frameworks, and they can be run on a wide range of computing platforms, including laptops, servers, and cloud-based systems.

Natural-Language Processing

Natural-language processing (NLP) is a field of AI and computer science that focuses on the development of algorithms and systems that can understand and process human language. This can include tasks such as language translation, text classification, and sentiment analysis.

To perform these tasks, NLP algorithms and systems use a variety of techniques, including machine learning, pattern recognition, and linguistics. These algorithms are trained on large datasets of labeled text, and they use statistical techniques to identify patterns and relationships within the data. Once trained, these algorithms can then be used to analyze and understand new text.

NLP algorithms and systems are widely used in a variety of applications, including chatbots, language-translation services, and social media analysis. These algorithms can be implemented using a variety of programming languages and frameworks, and they can be run on a wide range of computing platforms, including laptops, servers, and cloud-based systems.

NLP was initially developed as a powerful enabler for assistive technologies, for people with visual, speech, hearing, motor or cognitive disabilities. It can be used to develop assistive technologies for people with disabilities in a number of ways. For example, NLP can be used to develop text-to-speech systems that can convert written text into spoken language, which can be helpful for people who are blind or have compromised vision. NLP can also be used to develop speech-to-text systems that can convert spoken language into written text, which can be helpful for people who are unable to type or have difficulty with fine motor skills.

In addition to these applications, NLP can also be used to develop technologies that can understand and respond to spoken or written commands, which can be helpful for people who have difficulty using a mouse or keyboard. For example, an NLP-based system might be able to understand and respond to voice commands to open and close applications, navigate websites, or send emails.

Overall, NLP has the potential to significantly improve the lives of people with disabilities by providing them with more flexible and convenient ways to interact with computers and other devices. This tool is now more broadly used around the world and is behind many applications, like our email spam detectors, Google Translate, chatbots, and even our AI assistants, like Alexa and Siri.

And there you have it. All these algorithmic tools come together to power artificial-intelligence systems. So in summary, algorithms are a set of instructions that develop artificial intelligences, AIs. AIs use algorithmic tools—like machine learning, deep learning, neural networks, computer vision, and natural-language processing—to give computer systems the ability to emulate human thought and perform tasks without the need for human interaction.

Let's explore a few concrete examples of how algorithms show up in our world today.

Call to Action

To extend your learning, let's conduct an algorithm-exploration activity. This will help you gain a deeper understanding of algorithms and their real-world applications. This activity also encourages critical thinking and discussion around the ethical considerations of AI and algorithms.

1. Select one of the following algorithmic tools discussed in the chapter:

   a. Machine Learning

   b. Deep Learning

   c. Neural Networks

   d. Computer Vision

   e. Natural-Language Processing (NLP)

2. Research the Chosen Algorithm:

   a. Look up additional information about the chosen algorithm online or in textbooks.

   b. Understand its fundamental concepts, how it works, and its main applications.

3. Identify Real-World Applications:

   Find at least two real-world examples in which the chosen algorithm is used. These could be applications you use daily or have heard of in the news.

4. Share and Discuss:

    a. If you're part of a group or class, share your presentation with your peers.

    b. Discuss the significance of the chosen algorithm in shaping AI and its potential impact on various industries and society.

5. Reflect and Connect:

    a. Reflect on how your chosen algorithm contributes to the AI landscape and the broader implications it has on our daily lives.

    b. Consider the ethical and societal aspects associated with the use of this algorithm.

Reflection: Create a Learning Hero Journal Entry

1. Write a brief journal entry summarizing your key takeaways from this chapter.

_____

_____

_____

_____

_____

_____

_____

_____

_____

_____

2. Reflect on any new perspectives or insights you gained about algorithms and AI.

Chapter 3

The Light Side of Algorithms

Every time you pick up your smartphone, you are using algorithms. From using facial recognition or your thumbprint to unlock your phone, to finding the best route to your favorite restaurant with a GPS navigation system, algorithms are driving the world as we know it today. Currently, algorithms are present in everything that involves the digital world and even in some things that have been used for decades. For example, the calculator has been around for a long time and is a great example of the use of algorithms.

Today, algorithms have the potential to solve complex problems effectively, and at scale. They are increasingly assisting or replacing humans in making important decisions. They help decide who gets hired, how much to charge for insurance, and who gets approved for a mortgage or a credit card. In some cases, they also inform choices about sentencing, parole, and bail for individuals involved in the justice system.

This is what we call "the light side of algorithms," or the numerous potentially positive consequences of their use. Here are some examples of the light side of algorithms and AI:

1. Increased efficiency and productivity: Algorithms and AI systems can perform tasks faster and more accurately than humans.

2. Ability to perform tasks that are difficult or impossible for humans: Algorithms and AI systems can be used when data collection or analysis is too complex or time-consuming for humans, such as analyzing large amounts of data or processing images and videos.

3. Potential to solve complex problems: Algorithms and AI systems can be used to analyze and calculate complex issues that are difficult for humans to understand or tackle, such as predicting weather patterns or finding new drugs to treat diseases.

4. More accurate and objective decision-making: Algorithms and AI systems are not subject to the same biases or limitations as humans when processing data.

It is important to recognize and harness the potential benefits of algorithms and AI in order to maximize their positive impact on society. Here are a few real-world examples of how algorithms show up in technology today:

1. Search engines: Algorithms are used to rank and organize search results based on relevance and other factors. Search engines work by using algorithms to scan the internet for web pages that contain specific key words or phrases entered by users in search bars. Search engines use complex algorithms to find relevant web pages and rank them based on their relevance and other factors. This enables users to quickly and easily find the information they are looking for. There are several steps involved in this process:

   a. Crawling: The search engine uses automated programs called "crawlers" or "spiders" to visit web pages and follow links on those pages to discover new content.

   b. Indexing: As the crawlers discover new web pages, they add them to the search engine's "index," which is a massive database of all the web pages that the search engine has discovered.

   c. Ranking: When a user enters a search query, the search engine's algorithms search the index for web pages that are relevant to the query and that contain the key words or phrases that were entered. The search engine then ranks these web pages based on their relevance and other factors, such as the quality and credibility of the web page.

d. Displaying search results: The search engine displays the search results, usually in the form of a list of web pages, links to them, and a brief description of each. The search results are typically listed with the most relevant and highest quality web pages appearing at the top of the list.

2. Social media: Algorithms are used to personalize content recommendations, determine which posts to show in a user's feed, and identify fake or spam accounts. Some of the ways that social media platforms use algorithms and AI to drive engagement include:

a. Personalization: Social media platforms use algorithms to personalize the content that users see in their feeds. This can involve showing users content that is more relevant to their interests, as well as content from their friends and family.

b. Recommendation: Social media platforms also use algorithms to recommend content to users in whom they might be interested. This can involve recommending new accounts to follow or suggesting similar content to what a user has liked or shared in the past.

c. Moderation: Social media platforms use AI to help moderate content. This can involve using machine-learning algorithms to identify and remove spam or inappropriate content, or to flag content for human review.

d. Advertising: Social media platforms use algorithms to target specific users for ads based on their interests and demographics. This can involve using machine-learning algorithms to analyze users' past interactions with content and to predict what types of ads they might be interested in.

3. Online advertising: Algorithms and AI are used to target specific users based on their interests, location, demographics, and other factors. There are several steps involved in this process:

a. Advertisers create ads: Advertisers create ads that they want to display to users, and specify the target audience based on factors such as age, gender, location, interests, and more.

b. Advertisers bid on ad space: Advertisers then bid on ad space on websites and other platforms. The amount that advertisers are willing to pay for ad space is determined by a variety of factors, including the expected effectiveness of the ad and the target audience.

c. Ad exchanges match ads to users: Ad exchanges are online marketplaces where advertisers can buy and sell ad space. When a user visits a website or other platform that displays ads, the ad exchange uses algorithms and AI to match the user with relevant ads based on the users' interests, demographics, and other factors.

d. Ads are displayed to users: Once a match has been made between an ad and a user, the ad is displayed to the user on the website or platform that they are visiting. The ad may be in the form of a banner ad, a video ad, or another type of ad.

4. E-commerce: E-commerce websites use AI and algorithms in a number of ways to improve the user experience and to

drive sales. They personalize the user experience, optimize prices, detect fraud, and provide customer service, among other things. These technologies help to improve the efficiency and effectiveness of e-commerce sites and drive sales. Some of the ways they use AI and algorithms include:

a. Personalization: E-commerce sites personalize the content and product recommendations that users see on their sites. This can involve showing users products that are more relevant to their interests, as well as products that are similar to ones that they have viewed or purchased in the past.

b. Price optimization: E-commerce sites optimize product prices based on a variety of factors, such as demand, competitors' prices, and cost of goods. This can involve using machine-learning algorithms to analyze historical data and make predictions about future demand and pricing.

c. Fraud detection: E-commerce sites use AI to detect fraudulent activity, such as suspicious purchases or the use of stolen credit card numbers. This can involve using machine-learning algorithms to analyze patterns of behavior and to flag suspicious activity for further investigation.

d. Customer service: E-commerce sites may use AI-powered chatbots to provide customer service, such as answering questions or helping users navigate the site. These chatbots can use natural-language processing (NLP) algorithms to understand and respond to user inquiries.

5. Email filtering: Email-filtering uses artificial intelligence (AI) and algorithms to automatically classify and categorize emails using machine-learning algorithms to analyze the content and characteristics of emails, and to classify them into appropriate categories. This enables users to more easily manage their emails and to identify and filter out unwanted messages. There are several steps involved in this process:

a. Collection of data: To train an email-filtering system, a large dataset of emails is collected and labeled with their appropriate categories (for example, spam, promotion, personal, etc.).

b. Training of the model: The email-filtering system uses this labeled dataset to train a machine-learning model to recognize the characteristics of different types of emails. This involves feeding the model a large number of labeled emails and adjusting the model's parameters until it can accurately predict the categories of new, unseen emails.

c. Classification of emails: Once the model has been trained, it can be used to classify new emails as they are received. The model analyzes the content and other characteristics of the email—such as the sender, subject, and attachments—and uses this information to predict the email's category.

d. Filtering of emails: Based on the predicted categories, the email-filtering system can then route the emails to the appropriate folders or take other actions, such as marking emails as spam or deleting them.

6. Fraud detection: Fraud detection systems use artificial intelligence and algorithms to analyze the characteristics of transactions and automatically identify patterns and anomalies that may indicate

fraudulent activity. This enables organizations to more effectively detect and prevent fraudulent activity, such as in finance and healthcare industries. There are several steps involved in this process:

a. Collection of data: To train a fraud-detection system, a large dataset of transactions is collected, including both fraudulent and legitimate transactions. This dataset is labeled with information about whether each transaction was fraudulent or not.

b. Training of the model: The fraud-detection system uses this labeled dataset to train a machine-learning model to recognize the characteristics of fraudulent transactions. This involves feeding the model a large number of labeled transactions and adjusting the model's parameters until it can accurately predict the likelihood that new, unseen transactions are fraudulent.

c. Analysis of transactions: Once the model has been trained, it can be used to analyze new transactions as they occur. The model analyzes the characteristics of each transaction—such as amount, location, and parties involved—and uses this information to predict the likelihood that the transaction is fraudulent.

d. Detection of fraudulent transactions: Based on the predicted likelihood of fraud, the fraud-detection system can then flag transactions for further investigation or take other actions, such as blocking the transaction or alerting security personnel.

7. Predictive analytics: Predictive analytics is a type of data analysis that uses artificial intelligence and algorithms to make predictions about future outcomes. This is useful in a variety of fields, including marketing, finance, and health care. These predictions can be used to inform decision-making and to take proactive actions in a variety of contexts. There are several steps involved in this process:

a. Collection of data: To perform predictive analytics, a large dataset is collected that includes a range of variables that may be relevant to the prediction. This dataset may include historical data as well as real-time data.

b. Training of the model: A machine-learning model is then trained on the dataset to recognize patterns and correlations between the variables. This involves feeding the model a large number of labeled examples and adjusting the model's parameters until it can accurately make predictions about new, unseen data.

c. Analysis of data: The model is then used to analyze the data and make predictions about future outcomes. This can involve predicting the likelihood of certain events occurring, such as the likelihood of a customer churning or of a machine breaking down.

d. Use of predictions: The predictions made by the model can then be used to inform decision-making and to take proactive actions. For example, if a predictive-analytics model predicts that a customer is likely to churn, the business may take steps to retain the customer, such as offering them a discount or providing them with additional support.

8. Virtual Assistants: Whether you use Alexa, Siri, Google Assistant, or Cortana, AI assistants have become very common. These applications use AI to understand and respond to voice commands. These virtual assistants are integrated into a

range of devices, including smartphones, tablets, and smart speakers, and they can be used to perform a wide range of tasks, including setting alarms, playing music, answering questions, and controlling other smart devices. There are several steps involved in the process of how virtual assistants work:

a. Collection of data: Virtual assistants are trained on large datasets of human language and conversations in order to learn how to understand and respond to user input. This dataset may include transcriptions of real conversations, as well as labeled examples of specific types of inputs and responses.

b. Training of the model: A machine-learning model is then trained on this dataset to recognize patterns and correlations in the language data. This involves adjusting the model's parameters until it can accurately understand and respond to new, unseen inputs.

c. Processing of user input: When a user speaks to or texts a virtual assistant, the assistant's algorithms analyze the user's input and attempt to understand the meaning and intent behind it. This may involve using natural-language processing (NLP) algorithms to analyze the structure and meaning of the words and phrases used.

d. Generation of a response: Based on its understanding of the user's input, the virtual assistant generates a response. This may involve retrieving information from a database, performing a task or action, or generating a simple conversation.

9. Image Recognition and Photo Tagging: Another use for AI is image recognition and photo tagging. This technology is able to identify places, people, objects, buildings, and essentially any other variable in images. Image recognition allows for photo tagging to happen, which is essentially tagging and labeling images with key words that refer to the image's content, making it more searchable and easier to find. Programs like Google Photos use a combination of AI and machine-learning algorithms to perform image recognition and photo tagging. These algorithms are designed to analyze and understand the content of photos and to identify and classify objects, people, and

other subjects within the photos. There are several steps involved in this process:

a. Collection of data: To train an image-recognition or photo-tagging system, a large dataset of labeled images is collected. This dataset includes images that are labeled with the objects, people, and other elements that appear in them.

b. Training of the model: The image-recognition or photo-tagging system uses this labeled dataset to train a machine-learning model to recognize the characteristics of different types of objects and elements. This involves feeding the model a large number of labeled images and adjusting the model's parameters until it can accurately identify the objects and elements in new, unseen images.

c. Analysis of images: Once the model has been trained, it can be used to analyze new images and identify the objects and elements that appear in them. The model analyzes the content and characteristics of the image—such as the shapes, colors, and textures of the objects and elements—and uses this information to make predictions about what is depicted in the image.

d. Tagging of images: Based on its predictions, the image-recognition or photo-tagging system can then tag the images with labels indicating the objects and elements that appear in them. These tags can be used to organize and search through large collections of images.

10. Ride-sharing technology: Ride-sharing companies like Uber and Lyft would not be possible without the use of AI. They use a combination of AI and algorithms to match riders with drivers, optimize routes, and predict demand. When a rider requests a ride through the app, the service's algorithms match the rider with a nearby driver, based on factors such as their location, rating, and availability. The use of AI and algorithms is an integral part of the ride-sharing experience, helping to ensure that riders are matched with drivers efficiently, and that rides are completed as quickly and smoothly as possible. Some of the ways that ride-sharing technology uses AI and algorithms include:

a. Matching riders with drivers: When a rider requests a ride through a ride-sharing app, the app's algorithms match the rider with a nearby driver based on factors such as their location, rating, and availability. The app also takes into account the estimated time it will take for the driver to reach the rider, and the estimated time it will take to complete the trip.

b. Optimizing routes: Once a match has been made, the app's algorithms may also be used to optimize the route that the driver takes to pick up the rider and to arrive at the final destination. This can involve taking into account real-time traffic data, as well as the location and availability of other drivers.

c. Predicting demand: In addition to these real-time functions, ride-sharing technology also uses AI and algorithms to predict future demand for rides in specific areas, and to optimize the number of drivers that are on the road in anticipation of this demand. This can involve using machine-learning algorithms to analyze historical data on ridership and other factors.

11. Robots: Robots, like the Boston Dynamics robot, use a combination of artificial intelligence and algorithms to perform tasks and functions. These robots can navigate through space while they avoid obstacles, even if they don't know what obstacles might appear in their way. The Boston Dynamics Atlas robot is a highly advanced, humanoid robot designed to perform a wide range of tasks. There are several key technologies that enable it and robots like it to function:

a. Sensors: Robots are equipped with a range of sensors that allow them to perceive and understand their environment. These may include sensors like lidar, stereo cameras, and inertial measurement units, which allow the robot to build a detailed 3D map of its surroundings and to detect objects and obstacles.

b. Machine-learning algorithms: Robots use machine-learning algorithms to process and analyze the data from their sensors in real time, and to make decisions about how to move and act based on this information. For example, the Boston Dynamics robot uses machine-learning algorithms to plan and execute movements and actions, such as walking over uneven terrain or manipulating objects.

c. Control algorithms: Robots also use control algorithms to maintain their stability and to coordinate their movements. For example, the Boston Dynamics robot uses advanced control algorithms to balance itself and to adjust its movements in response to external forces and changes in its environment.

As you can see, there are so many uses for the powerful algorithms that rule our lives and organizations today.

AIs have the ability to change, adapt, and grow based on new data, which is why they are described as "intelligent." And this is where the problem with algorithms comes in—there is the possibility that the system will be fed fallible data, and as it learns, it amplifies possible mistakes. In turn, these mistakes are compounded when they lead to erroneous, harmful predictions.

I just shared with you a few examples of how computer algorithms show up in our lives today. Can you identify five more examples, from your daily life, in which you are using algorithms?

Call to Action

Take some time to identify the algorithms used in your daily life. Recognize and understand the prevalence of these algorithms and how they impact various aspects of your daily routine.

1. **Reflect on Your Daily Routine:** Take a moment to think about your typical day. Consider the various activities and interactions you have with technology and digital systems.

2. **Identify Instances of Algorithm Use:** For each activity or interaction, try to identify whether algorithms are involved. Algorithms may be responsible for decision-making, automation, personalization, or optimization. Here are some prompts to get you started:

   a. Smartphone use: Think about the apps you use, such as social media, navigation, or personal assistants. How do algorithms influence your experience with these apps?

   b. Online shopping: When you shop online, how are algorithms used to recommend products or personalize your shopping experience?

   c. Streaming services: How do algorithms play a role in suggesting movies, TV shows, or music?

d. Email: Consider how email-filtering algorithms categorize your emails into different folders, including spam.

e. Ride-sharing or food-delivery apps: How do these apps use algorithms to match drivers or delivery partners with customers and optimize routes?

f. Document Your Findings: Create a list or a diagram that outlines the activities you identified and the corresponding algorithms at play. Describe briefly how each algorithm impacts your experience or decision-making in that activity.

3. **Share and Discuss:** If you're participating in a group or class, share your findings with your peers. Discuss how algorithms have become integral to daily life and how they affect your routines and choices.

4. **Consider Ethical Implications:** Reflect on the potential ethical implications of algorithm use in these contexts. Are there situations in which algorithmic decisions may raise concerns about fairness, bias, or privacy?

Reflection: Create a Learning Journal Entry

1. Write a brief journal entry summarizing your key takeaways from this chapter.

_____

_____

_____

_____

_____

_____

_____

_____

_____

_____

_____

2. Reflect on any new perspectives or insights you gained about algorithms and AI.

Chapter 4

Trials and Tribulations: The Cost of Algorithms

Algorithms have given birth to some of the largest corporate empires—such as Google, Apple, Amazon, and Microsoft—and have helped businesses become successful by taking large amounts of data and making sense of it, which improved their decision-making processes and, eventually, their overall business performance. Prior to the use of algorithms, companies used to hire teams of data analysts to find patterns and create data trends for use in decision-making. Calculations that used to take weeks are now run using algorithms and are completed in a matter of seconds. Also, businesses are now able to shape demand for their products and services by collecting information about customers and defining unique experiences based on this information.

Even at the local government level, for instance, where effective trash collection is a manual task that is important to keep citizens happy, an algorithm could sift through all sorts of data—such as weather, time, traffic, and available resources—to determine the best possible route for trash pickup.

Algorithms also have the potential to save lives! For example, a hospital in Texas has been implementing algorithms to inform doctors which patients are at higher risk for heart failure. To do this, organizations are collecting a lot of personal data or personally identifiable information. For clarity, when you use computer systems, you are more than just a user or customer of these systems. You are a generator of data; the more that you use the system, the more actionable data you produce.

But as with any instrument or tool that can be operated effectively and safely, there is also the possibility of accidents resulting in danger and harm. There are a number of costs associated with the collection of data for AI algorithms, including:

1. Time and resources: Collecting and preparing data for use in AI algorithms can be a time- and resource-intensive process, particularly if the data is large or complex. This can involve costs for data scientists and other professionals to collect, clean, and label the data, as well as for hardware and software resources to process and store the data.

2. Privacy and security: If data is being collected from individuals, there may be costs associated with obtaining consent

and ensuring that the data is collected and used in a way that respects privacy and security. This can involve costs for legal and compliance professionals, as well as for systems and processes to protect the data.

3. Opportunity cost: There may also be opportunity costs associated with data collection, such as the opportunity cost of using the resources and time spent on data collection for other purposes.

The costs of data collection for AI algorithms can vary depending on the specific requirements and use cases, but it can be a significant investment for organizations that rely on this data. Not to mention that we, as individuals, are signing away certain rights when we accept end-user license agreements on social media platforms. If we think about the process, it becomes clear that we are being asked to agree to something that we most likely would not under any other circumstances, something that we probably don't even fully understand.

Our very lives, our locations, our family members, our preferences, our dislikes—all of this is not really actionable data until you create algorithms that find patterns and convert every single human being's actions into a collection of data that collectively can be used to exploit us. Some people even say that this gathering of our personal data makes us, as humans, hackable.

There are a number of ethical and moral questions that individuals and society should consider when it comes to data collection for algorithms. Here is a brief checklist:

1. Who has access to the data?

2. Is the data being collected and used by a single organization, or is it being shared with third parties?

3. How is the data being used?

4. Is the data being used for the purpose for which it was collected, or is it being used for other purposes?

5. What are the implications of the data being used in this way?

6. Could the data be used to discriminate against certain individuals or groups, or to have other negative consequences?

7. Is the data being collected with the consent of the individuals whose data it is?

8. Do individuals understand how their data is being collected and how it will be used?

9. Is the data being collected and used in a way that respects the individuals' privacy?

10. Are appropriate measures in place to protect the security and confidentiality of the data?

11. Is the purpose of the data collection clear and transparent?

12. Do individuals have the ability to opt out of the data collection if they choose?

13. Is the data being collected and used in a way that is fair and unbiased?

As we all know, at an unprecedented pace, technology is changing the way we live and work. When the cost of algorithms leads to a lack of trust in algorithms, it can have a number of negative consequences, including:

1. Loss of confidence: If individuals lose trust in algorithms, they may be less likely to rely on them or to trust the decisions and outcomes that they generate. This could lead to a decline in the use of algorithms, or to a mistrust of the technology more broadly.

2. Disruption of systems and processes: If algorithms are widely used to make decisions and to drive certain processes, a loss of trust in these algorithms could lead to disruptions in these systems and processes. This could have far-reaching consequences, depending on the nature of the algorithms and the systems they are used in.

3. Damage to reputation: If trust in algorithms is broken, it could also lead to damage to

the reputation of the organizations that use the algorithms, as well as to the reputation of the technology more broadly.

Let's transition into revealing more about the dark side of algorithms and why your role as a leader is so important in ensuring that technology is used in a way that is inclusive and benefits everyone.

As you venture further on your AI hero's journey, you'll encounter ethical dilemmas that demand your courage and integrity. We explore the complexities of AI ethics and guide you in making informed decisions that prioritize human welfare and respect for privacy. Engaging case studies and ethical frameworks will help you navigate through the stormy seas of AI with grace and responsibility.

The Dark Side of Algorithms

Trust is the foundation for any strong society. It is the ultimate currency in the relationships on which all communities are built—and not just with our family and friends—but also with businesses, governments, and the media. We trust banks with our money. We trust doctors with our personal information.

As more of our daily decisions are made by algorithms, it is important to have trust in these algorithms, as well, in order for them to be effective and to have a positive impact on our lives. There is a risk that trust in algorithms could be broken if the algorithms are found to be flawed, biased, or unethical in some way. What happens when this trust is broken?

That is where the dark side of algorithms comes in. The dark side of algorithms includes negative consequences such as amplification of bias, erosion of privacy, and loss of accountability. It is important to ensure that algorithms are transparent, unbiased, and ethical in order to maintain trust in them.

There are a number of potential negative consequences or "dark sides" to the use of algorithms, including:

1. Bias: Algorithms are only as unbiased as the data and the human biases that go into designing and programming them. As a result, algorithms can sometimes perpetuate or amplify existing biases, which can lead to outcomes that are unfair or discriminatory.

2. Transparency: Algorithms can be complex and hard to understand, even for the people who design and program them. This lack of transparency can make it difficult for individuals to understand how and why certain decisions or outcomes are being generated, which can lead to a lack of accountability and trust.

3. Loss of privacy: The use of algorithms often involves the collection and analysis of large amounts of data, which can raise concerns about privacy. If this data is not properly secured, or is used in ways that individuals do not understand or consent to, it can lead to a loss of privacy and a sense of loss of control over personal information.

4. Dependence: As algorithms become increasingly prevalent, there is a risk that individuals and organizations will become overly reliant on them, which can lead to a loss of critical thinking and decision-making skills. This dependence on algorithms can also lead to a lack of flexibility and adaptability in the face of changing circumstances.

It is important to be aware of these and other potential negative consequences of the use of algorithms, and to take steps to mitigate these risks, in order to ensure that algorithms are used in a responsible and ethical manner. Let's look at each of these consequences in more detail.

Bias in Algorithms

Bias in algorithms and artificial intelligence (AI) refers to the systematic pattern of deviation from the norm, or from the expected value, which results in certain outcomes being more or less likely. Bias can occur at various stages of the development and implementation of algorithms and AI systems, and it can have negative consequences for the individuals or groups affected by it.

There are many potential sources of bias in algorithms and AI, including:

1. Data bias: This occurs when the data used to train an algorithm or AI system is unbalanced or unrepresentative of the population it is intended to serve. For example, an algorithm that is trained on data from predominantly White, male job applicants may exhibit bias against female and minority applicants.

2. Algorithmic bias: This occurs when an algorithm is designed or trained in a way that leads to biased outcomes. For example, an algorithm that is designed to predict the likelihood of loan default may be biased against certain groups if it is trained on data that reflects historical patterns of discrimination.

3. Human bias: This occurs when humans, who may have their own biases, design or implement algorithms or AI systems in a way that leads to biased outcomes. For example, a human may design an algorithm that is intended to identify job candidates who are team players, but the

criteria they use to define this may be biased against certain groups.

4. Interactional bias: This occurs when the interaction between an algorithm or AI system and the people it serves is biased. For example, an AI-powered hiring tool that is intended to assist recruiters in identifying the best candidates for a job may exhibit bias if it is more likely to present candidates from certain groups to recruiters.

It is important to recognize and address bias in algorithms and AI systems, as it can have significant consequences for the individuals and groups affected by it.

Transparency or Opacity in Algorithms

Opacity or transparency in algorithms and artificial intelligence refers to the degree to which the inner workings of the algorithm or AI system are open to inspection and understanding. A transparent algorithm is one that is open and clear in its design, function, and decision-making processes, while an opaque algorithm is one that is difficult or impossible to understand or explain.

Here are several examples of both transparency and opacity in algorithms and AI:

1. Transparent algorithms: These are algorithms that are designed and implemented

in a way that allows their inner workings and decision-making processes to be understood and inspected. For example, a transparent machine-learning algorithm may allow users to see the features and variables that were used to make predictions, as well as the model's accuracy and error rate.

2. Opaque algorithms: These are algorithms that are difficult or impossible to understand or explain, often because they are complex or use proprietary techniques. For example, an opaque AI system may be based on deep-learning techniques that are too complex to be understood by humans, or it may be protected by intellectual property laws that prevent others from examining its inner workings.

Transparency is important in algorithms and AI because it allows collaborators to understand how the system works and how it makes decisions. This is particularly important when the system is being used to make decisions that have significant consequences for individuals or groups, as it allows for accountability and fairness. Opacity, on the other hand, can make it difficult or impossible to understand how an algorithm or AI system is making decisions, which can lead to mistrust and skepticism.

Loss of Privacy in Algorithms

Loss of privacy in algorithms and artificial intelligence refers to the risk that personal information or data may be collected, used, or shared in ways that are outside the control or knowledge of the individual. Here are several ways in which the use of algorithms and AI can lead to a loss of privacy:

1. Data collection: Algorithms and AI systems often rely on the collection of large amounts of data to function properly. This data may include personal information such as names, addresses, and contact information, as well as more sensitive information such as medical records or financial transactions. If this data is not properly secured or protected, it can be accessed by unauthorized parties and used in ways that compromise an individual's privacy.

2. Data sharing: Algorithms and AI systems may be designed to share data with third parties, such as marketers or advertisers, in order to improve performance or generate revenue. This data sharing can occur without the individual's knowledge or consent, and it may expose personal information to parties who may use it in ways that are outside the individual's control.

3. Data mining: Algorithms and AI systems may be used to analyze and mine data for insights or patterns that are not immediately apparent to humans. This data mining can reveal sensitive information about an individual's habits, preferences, or behavior, and it may be used to make decisions about that individual without their knowledge or consent.

4. Predictive analytics: Algorithms and AI systems may be used to make predictions about an individual's future behavior or outcomes based on their past data. This can include predictions about an individual's likelihood of making certain choices or experiencing certain events, and it can be used to make decisions about that individual without their knowledge or consent.

It is important to be aware of the potential risks to privacy that can arise from the use of algorithms and AI, and to take steps to protect personal information and data. This may include measures such as encrypting data, limiting data collection and sharing, and being transparent about how data is used.

Dependence on Algorithms

Dependence on algorithms and artificial intelligence refers to the reliance on them to perform tasks or make decisions, to the extent that it becomes difficult or impossible to function without them. There are several ways in which dependence on algorithms and AI can occur, making it difficult or impossible to function without them:

1. Automation: Algorithms and AI systems may be used to automate tasks that were previously performed by humans, such as data entry or customer service.

2. Decision-making: Algorithms and AI systems may be used to make decisions that were previously made by humans, such as loan approvals or hiring decisions.

3. Information processing: Algorithms and AI systems may be used to process and analyze large amounts of data in order to extract insights or make predictions.

4. Infrastructure: Algorithms and AI systems may be integrated into the infrastructure of an organization or society, such as transportation systems or power grids.

Dependence on algorithms and AI can have both positive and negative consequences. On the one hand, it can lead to increased efficiency and productivity, as algorithms and AI systems can perform tasks faster and more accurately than humans. On the other hand, it can also lead to a loss of human jobs and a reliance on technology that may be vulnerable to failure or abuse.

It is important to carefully consider the potential consequences of dependence on algorithms and AI, and to take steps to mitigate potential negative effects.

The dark side of algorithms refers to the potential negative consequences of their use, including the amplification of bias, the erosion of privacy, and the loss of accountability. These negative consequences can have significant impacts on individuals and society, and it is important to recognize and address them in order to ensure that the benefits of algorithms and AI are realized in a responsible and ethical manner.

Examples of the Dark Side of Algorithms

While the tech industry may be full of developers who believe in unleashing creativity and innovation, there are also those who intentionally or unintentionally harm society by, for example, monitoring customers and invading privacy, developing systems that offer women less credit than men, making it harder for people with mental health issues to get jobs, or treating minority defendants more harshly than White ones.

Here are some examples of the negative consequences of algorithms and artificial intelligence, often referred to as the "dark side" of algorithms:

- Amplification of bias: Algorithms and AI systems can perpetuate and amplify existing biases in the data they are trained on, leading to unfair and discriminatory outcomes. For example, an algorithm that is trained on data from predominantly White-male job applicants may exhibit bias against female and minority applicants.

- Erosion of privacy: The collection and use of personal data by algorithms and AI systems can compromise individuals' privacy and expose them to risks such as identity theft and unauthorized data sharing.

- Loss of accountability: The use of algorithms and AI systems to make decisions that have significant consequences for individuals or society can make it difficult to hold those responsible for the outcomes accountable. This can lead to a lack of transparency and a sense of unfairness when things go wrong.

- Unemployment: The automation of tasks by algorithms and AI systems can lead to the displacement of human workers, which can have negative consequences for employment and income.

- Security risks: Algorithms and AI systems may be vulnerable to cyberattacks or other security threats, which can have significant consequences for individuals, organizations, and society.

- Dependence: The reliance on algorithms and AI systems to perform tasks or make decisions can lead to a dependence on technology that may be vulnerable to failure or abuse.

In addition to defining the consequences, it is important to recognize and address real-world examples of the negative consequences of algorithms and AI in order to ensure that they are used in a responsible and ethical manner. Here are some examples to illustrate the implications and consequences of the dark side of algorithms.

Skin Tone and Algorithms

AI bias with regards to skin tone refers to the systematic patterns of deviation from the norm or expected value that result in certain outcomes being more or less likely for individuals with different skin tones. AI bias can occur when algorithms and AI systems are trained on data that is unbalanced or unrepresentative of the population it is intended to serve, and this can lead to unfair and discriminatory outcomes for certain groups.

There have been a number of incidents in which AI systems have been found to exhibit bias with regards to skin tone:

- In 2009, users of Hewlett-Packard webcams reported that webcams were unable to recognize faces of dark-skinned people. Unfortunately, this wasn't the last time that facial-recognition systems inadvertently maligned dark-skinned people.

- Four years later, in 2013, Flickr, an online photo-management and sharing platform, introduced an auto-tagging feature that used artificial intelligence to automatically identify and label objects and people in photos. However, the feature was found to have labeled a photo of a Black person as an ape, causing outrage and accusations of racism.

- In 2015, Google Photos, a service that uses artificial intelligence to automatically tag and organize photos, was found to have labeled photos of dark-skinned individuals as gorillas. This caused outrage and accusations of racism, as the AI system had been trained on a dataset that did not include images of gorillas and therefore did not have any context for the concept. Google apologized for the incident and implemented changes to the algorithm to prevent similar incidents from occurring in the future.

These incidents illustrate the clear risks of bias in AI systems, as the algorithms are continuously being built to perpetuate and amplify biases that exist in the data on which they are trained. They also highlight the importance of ensuring that AI systems are trained on diverse and representative datasets in order to avoid such biases and to ensure that they are able to perform accurately and fairly.

Gender and Algorithms

Gender bias in AI is a specific form of AI bias that occurs when an AI system exhibits discriminatory behavior towards individuals based on their gender. This can manifest in a variety of ways, including through the AI system's output or decisions, or through the way it processes and interprets data.

There are several ways in which gender bias can be introduced into AI systems. For example, it can be introduced through the data that is used to train the system if that data is not representative of the population on which the system will be used. It can also be introduced through the way the system is designed and the algorithms it uses if that information incorporates biased assumptions, or if it is not designed to be fair and unbiased.

There are many examples of gender bias in algorithms. Here are a few:

- In 2015, it was discovered that the Beauty. AI algorithm, which was used to select winners for a global beauty contest, was biased against certain ethnicities and favored White participants.

- In 2016, it was found that the COMPAS algorithm, which was used to predict the likelihood of recidivism in criminal offenders, was biased against Black defendants and overestimated their risk of reoffending.

- In 2017, researchers found that a widely used machine-learning algorithm for predicting cardiovascular risk was less accurate for women than for men, due in part to the way the algorithm was trained on data that was predominantly from men.

In 2018, a study found that a machine-learning algorithm used to predict the likelihood of returning to work after a medical leave was biased against women, leading to lower predictions for women than for men with similar medical histories.

These are just a few examples of the gender bias in algorithms that has been identified. Addressing gender bias in AI is important because AI systems are increasingly being used in decision-making processes that can have significant impacts on individuals and communities. Bias in these systems can perpetuate and amplify existing inequalities and discrimination. To address gender bias in AI, it is important to identify and eliminate biases in the data used to train the system, as well as to design and test the system to ensure it is fair and unbiased.

Criminal Justice System and Algorithms

Algorithms can also determine the trajectory of your life. The criminal justice system is the system of practices and institutions by which a society deals with crime. It includes the police, courts, and corrections. Algorithms can be used in various aspects of the criminal justice system, including predicting crime, identifying suspects, determining sentences, and monitoring offenders.

One example of the use of algorithms in the criminal justice system is seen in risk-assessment tools. These tools use algorithms to predict the likelihood of an individual committing a crime in the future, based on data such as their criminal history, age, and other factors. This information can be used by judges to determine bail or sentencing.

There are concerns about the use of algorithms in the criminal justice system, particularly the potential for biased outcomes. For example, if the data used to train an algorithm is biased, the algorithm may produce biased results. There is also a concern that the use of algorithms may lead to a lack of transparency and accountability in decision-making.

ProPublica, an independent, nonprofit newsroom, launched an investigation into an algorithm used to decide whether a person charged with a crime should be released from jail prior to their trial. The journalists found that the Black people who did not go on to reoffend were assigned high risk scores, while the White people who did not go on to reoffend received low risk scores. Intuitively, the different false-positive rates suggest a clear-cut case of algorithmic racial bias.

Scores like these, known as risk assessments, are generated by computer algorithms and are increasingly being relied upon in courtrooms. They are used to inform decisions about who can be set free at every stage of the criminal justice system, from assigning bond amounts, to even more fundamental decisions about defendants' freedom. The results of such assessments are given to judges during criminal sentencing.

The use of algorithms in the criminal justice system is a complex and controversial issue. It is important to carefully consider the potential benefits and drawbacks of their use.

Facial Recognition and Algorithms

Facial-recognition systems are algorithms used to identify and compare faces in digital images. They have been used for a variety of applications, such as access control, surveillance, and crime prevention. However, recent studies have revealed that these algorithms can be biased against women and racial minorities. A 2012 study found that when used on photos of men, Whites, or the middle-aged, the best commercial systems matched faces successfully about 94.5 percent of the time, but their success rates were lower for women (89.5 percent), Blacks (88.7 percent), and the young (91.7 percent).

In one example of algorithmic bias, in 2020, Detroit police drove to the suburb of Farmington Hills and arrested Williams in his driveway while his wife and young daughters looked on. Williams, a Black man, was accused of stealing watches from Shinola, a luxury store. He was held overnight in jail.

Officers reportedly reviewed the surveillance-video image and saw that their facial-recognition technology had falsely identified Williams by matching him to the image of the suspect—an image that did not look like Williams. As you can see, facial-recognition technology poses serious threats to some fundamental human rights.

In fact, in 2019 a federal government study in the US confirmed that racial bias existed in many of their facial-recognition systems. Even their top-performing systems misidentified dark-skinned people at rates five to ten times higher than White-skinned people. The National Institute of Standards and Technology (NIST) released a study which found that 189 facial-recognition algorithms exhibited bias, most of which falsely identified Black and Asian faces ten to one hundred times more often than they did White faces. The study also found that the algorithms falsely identified women more than they did men, making Black women particularly vulnerable to algorithmic bias.

Furthermore, algorithms using US law-enforcement images falsely identified Native Americans more often than people from other demographics. These findings have prompted the introduction of the Algorithmic Accountability Act, which, if passed, would direct the Federal Trade Commission (FTC) to regulate the industry and require companies to assess their technology for fairness, bias, and privacy issues.

Fortunately, recent improvements in facial-recognition technology have led to a significant reduction in the error rates for these groups. These improvements are largely the result of a couple of technical issues, such as an expanded training set and improved lighting and lens exposures used to capture images. However, disparities may still exist in more difficult tests, such as "one-to-many" searches.

Discrimination in Online Ad Delivery

Online ad delivery refers to the process of delivering online advertisements to users through various channels, such as websites, social media platforms, and apps. There is a concern that online ad-delivery systems may discriminate against certain groups of users, based on characteristics such as race, gender, and age. This can occur through various mechanisms, such as the use of algorithms that are trained on biased data or the use of data that reflects existing societal biases.

For example, there have been reports that Black users were shown fewer job and housing ads than White users on some platforms. There have also been reports of gender-based discrimination in online ad delivery, with women being shown ads for lower paying jobs and men being shown ads for higher paying jobs.

Latanya Sweeney, a computer scientist and data privacy researcher, documented in a Harvard University paper that after Googling her name, the ads that appeared alongside her results suggested that she may have been arrested. The Google ad suggested that Latanya Sweeney had a criminal record, the details of which could be accessed by clicking on the ad. But after clicking on the ad and paying the necessary subscription fee, Sweeney says she found no record of arrest.

After collecting over 2,000 names and reviewing the results, Sweeney and her colleagues found that search ads for arrest records were more likely to appear when people searched for names that were more common among Black people, as compared to names that were more common among White people. This suggested that the ad-delivery system was discriminating against Black users based on their names.

Sweeney has also studied the use of personal data in online ad-delivery systems and its potential for use in discriminating against users. She argues that the use of personal data in online advertising should be regulated to protect the privacy and rights of users.

Discrimination in online ad delivery can have significant consequences for individuals and for society as a whole. It can lead to certain groups being disadvantaged and excluded from opportunities, and it can contribute to the perpetuation of societal inequalities. Sweeney's research helped to shed light on issues related to discrimination and online ad delivery, and it has contributed to the development of policies and practices to address these issues. It is important for online ad-delivery systems to be designed and implemented in a way that minimizes the risk of discrimination.

Racial Bias in Medical Algorithms

There is a concern that medical algorithms may be biased against certain racial groups and may contribute to health disparities. One example of this is the use of algorithms to predict the risk of a patient experiencing a medical condition such as heart disease or diabetes.

In an algorithm from the health services company Optum, health costs were used to predict and rank which patients would benefit most from extra care that could help them stay on their medications and keep them out of the hospital. But researchers in 2019 said that using health costs as a proxy for health needs is biased because Black patients face disproportionately higher levels of poverty and often spend less on health care than Whites.

If the data used to train these algorithms is biased, the algorithms may produce biased results that disproportionately affect certain racial groups. For example, if the data used to train an algorithm is predominantly from White patients, the algorithm may not accurately predict the risk for patients of other races, which can lead to unequal treatment. As a result of this bias, the algorithm used by Optum falsely concluded that Blacks were healthier than equally sick White patients.

There have also been concerns about the use of algorithms to assist in diagnoses and treatment decisions. If these algorithms are trained on biased data, they may produce biased recommendations that disproportionately affect certain racial groups.

It is important for medical algorithms to be designed and implemented in a way that minimizes the risk of bias and ensures that all patients receive fair and equal treatment. This may involve using diverse representative data to train the algorithms, as well as ongoing efforts to monitor and address any potential biases that may arise.

The examples are endless. They show how AI can foster discrimination, the lack of equal opportunity, and social exclusion. These biases are perpetuated because of app developers relying on unrepresentative training data or prejudiced historical data, or failing to address statistical biases. For example, algorithms used in artificial intelligence are often trained on datasets which reflect racial, gender, and other human biases. This can lead to algorithms exhibiting bias and disproportionately harming marginalized populations. Additionally, if developers do not address statistical biases, then the algorithms can be inaccurate and lead to false matches. To address these issues, developers need to ensure that their algorithms are equipped with accurate, diverse datasets and that bias is assessed and addressed during the development process.

As a leader, one of your roles in a technology company should be to ensure usability tests, explorations, and any development research include users from a range of backgrounds, household incomes, life stages, races, ethnicities, disabilities, sexes, genders, and more.

The more diverse the data, the richer the experience will be for all.

Call to Action

In this chapter, we've explored the potential negative consequences of algorithms and artificial intelligence, including bias, loss of privacy, and dependence. It's crucial to ensure that algorithms are used responsibly and ethically. To extend your learning and take concrete steps to develop ethical and unbiased algorithms, consider the following:

1. **Identify Algorithm Usage:** List the algorithms or AI systems currently used in your organization or community. Include their purposes and applications.

2. **Data Collection and Handling:** Assess how data is collected, stored, and used for these algorithms. Consider the following:

   a. Are there any privacy concerns related to data collection?

   b. Is data being shared with third parties?

   c. Is the data used in a way that respects individuals' privacy?

3. **Bias Evaluation:** Analyze the potential for bias in your algorithms. Think about:

   a. The diversity of the data used for training.

   b. Whether the algorithms perpetuate or amplify existing biases.

   c. Any known incidents or concerns related to bias.

4. **Transparency Assessment:** Examine the transparency or opacity of algorithms. Ask:

   a. Can the inner workings of these algorithms be understood and inspected?

b. Is there transparency in the decision-making process?

5. **Privacy Protection:** Evaluate the measures in place to protect individuals' privacy. Consider:

   a. Data encryption and security measures.

   b. Consent processes for data collection.

6. **Dependence and Accountability:** Reflect on the level of dependence on these algorithms. Ask:

   a. Are critical decisions heavily reliant on algorithms?

   b. Is there accountability in place for algorithm-generated outcomes?

   c. Mitigation and improvement: Identify areas in which improvements are needed to address ethical concerns. Develop strategies to mitigate risks, such as bias-reduction techniques or enhanced transparency measures.

7. **Community Engagement:** If applicable, involve your community or organization's stakeholders in the discussion about algorithms and their ethical implications. Seek input and feedback.

8. **Action Plan:** Develop an action plan that outlines steps to improve the ethical use of algorithms in your context. Set clear goals, responsibilities, and time lines.

9. **Regular Review:** Commit to regularly reviewing and updating your ethical algorithm assessment to adapt to changing circumstances and advancements in AI technology.

Reflection: Create a Learning Hero Journal Entry

1. Write a brief journal entry summarizing your key takeaways from this chapter.

_____

_____

_____

_____

_____

_____

_____

_____

_____

_____

_____

2. Reflect on any new perspectives or insights you gained about algorithms and AI.

Chapter 5

Activating the AI-Hero Superpowers

As the push for diversity in technology and the disruption of AI bias gains momentum, it's evident that the tech industry remains predominantly occupied by White males, which makes it an exclusive environment. This lack of inclusivity can be attributed to various factors, including a lack of diverse representation in leadership positions, a limited access to education and resources, and the perpetuation of unconscious biases and stereotypes.

Moreover, some technology companies have been hesitant to address or acknowledge the issue of diversity, hindering progress towards meaningful change. To foster a more inclusive tech industry, organizations must actively strive to recruit, hire, and retain diverse talent while cultivating a workplace culture that mitigates bias, embraces transparency, prioritizes privacy, and encourages inclusivity.

In addition to building diverse teams across the organization, one approach to initiate change is establishing an **algorithmic auditing cycle of inquiry** facilitated by an **ethical standards team**. This initiative acts as a checks-and-balances mechanism, identifying problems and providing opportunities to suggest improvements or alternatives to developers.

While many tech companies primarily focus on getting AI to function in production, there is a pressing need to ensure ethical standards. As a leader, your responsibility lies in establishing an ethical standards team comprised of developers, leaders, and non-technology team members, such as legal, HR, and product management. It is crucial that this team be composed of a cross section of diverse backgrounds and opinions in order to foster an ethics-driven process.

Once the team is formed, they can engage in critical inquiries such as, "How are we utilizing and securing the data to which we have access?" or "Does the data sample appropriately reflect the reality of our market?" These questions challenge existing distributions and prompt thoughtful decision-making to address biases and improve inclusivity.

You might be wondering about the practicality of assembling such a team and initiating these discussions. It's important to understand that the goal is not to be present in every room, but rather to equip your team members with the tools to recognize and interrupt biases, and to provide support when needed. This process holds immense significance as it underscores the inherent fallibility of even the most well-trained, experienced, and well-intentioned systems and designs. In our current landscape, there is a pressing need for a steadfast commitment to designing for all users and integrating inclusion into every decision we make.

By adopting an algorithmic auditing cycle of inquiry and seamlessly incorporating it into the development process, we can meticulously review code, promptly identify errors, and prevent potentially harmful violations. Conducting regular audits throughout the development journey saves valuable time and resources, enabling us to swiftly detect and resolve any issues before they escalate.

Let us not forget the essence of being an AI hero, as encapsulated by the acronym HERO. When we say:

"HONOR YOUR FOUNDATION"—we recognize and value the rich tapestry of diversity that exists within our global community. We acknowledge the unique experiences, perspectives, and contributions of individuals from all walks of life. By honoring our foundation, we create a culture that celebrates diversity and fosters an environment in which everyone's voice is heard and respected.

"EMBRACE EMPATHY"—we are reminded of the importance of understanding and empathizing with the experiences of others. It challenges us to actively seek out opportunities for growth and improvement in fostering equity and inclusivity. By embracing empathy, we strive to create spaces that are safe, welcoming, and free from prejudice, ones in which individuals are seen, heard, and valued for who they are.

"RESPECT IN ACTION"—we are called to go beyond passive recognition and to take tangible steps towards promoting kindness, solidarity, and understanding. It empowers us to stand up against injustice, to amplify marginalized voices, and to advocate for change. Through respect in action, we build bridges of understanding and work towards a more compassionate and inclusive society.

And finally, "OPTIMIZE INCLUSION"— we are compelled to engage in activism that supports a culture of belonging and works towards positive change. It urges us to challenge systemic barriers, promote equal access and opportunities, and dismantle structures that perpetuate inequality. By optimizing inclusion, we strive to create a world in which everyone can fully participate, thrive, and contribute their unique talents.

Let us remember that these principles serve as the framework for activating our AI-hero superpowers, which will enable us to shape a future in which innovation is rooted in diversity, empathy, respect, and inclusion. By embracing this powerful framework, we can truly become agents of change, leveraging the power of artificial intelligence to drive positive and transformative forces for change in our world. Together, we can create a more inclusive and equitable future in which AI systems empower and uplift individuals from all walks of life.

1. Honor Your Foundation: Develop a Foundational Inclusion Strategy

Before tackling the issue of algorithmic bias, it is important to develop a foundational inclusion strategy. This refers to creating a comprehensive plan that outlines the actions and goals that you or an organization will take to promote inclusion and be held accountable for their creations. The ideal strategy should address a range of issues, including promoting inclusive environment and addressing issues of equity and fairness. Creating an inclusive and belonging culture is of utmost importance for several reasons:

a. Enhancing Decision-Making and Innovation

By fostering a diverse workforce with a range of perspectives and experiences, organizations can tap into the collective wisdom that leads to better decision-making and innovative solutions.

b. Building A Good Reputation and Fostering Customer Trust

A strong commitment to inclusion and belonging can bolster a company's reputation and build trust with customers, especially among underrepresented groups who value diversity and fair representation.

c. Attracting and Retaining Top Talent

Companies that prioritize inclusion and belonging are more likely to attract and retain talented individuals, particularly those from underrepresented backgrounds who seek inclusive and equitable work environments.

d. Cultivating a Positive and Inclusive Culture

An inclusive strategy nurtures a positive work culture in which every team member feels valued, respected, and empowered. This, in turn, enhances employee satisfaction, engagement, and overall well-being.

e. Unlocking Market Opportunities

By authentically representing and meeting the needs of diverse customer groups, organizations with a strong inclusion focus can tap into new market opportunities and gain a competitive edge.

How to Establish an Inclusion Strategy for Your Success

To establish a robust foundation for an inclusive culture, consider the following key elements:

a. **Mission Statement:** Craft a clear and concise statement that articulates the organization's commitment to fostering an inclusive and belonging culture.

b. **Goals and Objectives:** Set specific, measurable, achievable, relevant, and time-bound (SMART) goals and objectives that outline what the organization aims to achieve in terms of inclusion and belonging.

c. **Key Actions:** Identify a comprehensive list of specific actions the organization will take to advance its inclusion and belonging goals, including initiatives such as training programs, mentorship opportunities, and diverse hiring practices.

d. **Metrics:** Establish measurable metrics to track progress and evaluate the effectiveness of the inclusion strategy, ensuring ongoing accountability and continuous improvement.

e. **Responsibility and Accountability:** Define clear roles and responsibilities for individuals and teams involved in implementing the inclusion strategy, and then establish accountability measures to ensure progress and success.

2. Embrace Empathy: Mitigate Bias and Cultivate Inclusion Equitably

The next vital stride is to proactively address biases and foster an inclusive environment that upholds respect and fairness for all. In this new era and new normal, a resounding chorus echoes, "Not on my watch." It is an unwavering refusal to tolerate injustice, a collective resolve to break the silence. Voices are amplified, actions resonate, and we are summoned to embrace this newfound sense of empowerment and agency.

Imagine the team you lead, the individuals who rely on your guidance and vision. They form the heartbeat of your community and organization. And their fulfillment and triumph lie at the essence of your aspirations. We, as humans, are united by a shared purpose, an invisible human-generated and -centered force that propels us all forward. It is what Abraham Maslow coined "self-actualization," the pursuit of reaching our fullest potential. As leaders, our mission is to cultivate an environment that nurtures this self-actualization, a space in which we and individuals we are blessed to serve can flourish and fulfill their unique missions.

The phrase "One purpose, many missions"—often attributed to Simon Sinek, an esteemed author, motivational speaker, and leadership expert—encapsulates the idea that while we share a common purpose or overarching goal in life, each person embarks on their own distinct set of missions or paths that contribute to the fulfillment of that shared purpose. It acknowledges the diversity of talents, passions, and aspirations among individuals, while recognizing that these missions, when aligned and harmonized, collectively contribute to the greater purpose we aspire to achieve as a collective.

In this era of transformative change, we are called upon to recognize the shifting paradigms of work and the emergence of a new generation. We must create an environment that transcends traditional notions, one in which every individual is cherished, empowered, and profoundly motivated.

Because here lies the truth: We are not driven by a singular mission alone. We are multifaceted beings with diverse talents, passions, and aspirations. By fostering an environment that embraces this truth, we enable our team members to explore their full potential, find purpose in their work, and make meaningful contributions, thereby increasing innovation and diversity of thought in our communities.

As leaders, our responsibility is to cultivate this extraordinary workplace in which the pursuit of excellence and the fulfillment of individual missions intertwine. It is about creating a culture that values collaboration, fosters growth, and celebrates the unique contributions of each person. When we unlock the full potential of our teams, we not only create an environment in which people thrive, but we also unleash a wellspring of creativity and innovation that propels our organization forward. It is a symbiotic relationship in which individuals are empowered to make a difference, and the collective brilliance of the team propels the organization to new heights.

Inclusion efforts have long involved investing in training, but the outcomes often fall short. It's not that training itself is ineffective; rather, knee-jerk reactions lead to lackluster, one-size-fits-all programs. This approach brings challenges, as research highlights. Mandated training irrelevant to individuals invites backlash and defensiveness. For underrepresented groups, it feels disingenuous.

As an AI hero, championing high-quality, targeted training programs is vital. All departments, from marketing to customer service, should embrace diversity, equity, and inclusion. By integrating these principles into existing initiatives, we create a holistic approach that fosters and then resonates with meaningful change.

Here are actionable measures an organization can undertake to support team members in their efforts to interrupt their biases and seek necessary assistance:

a. Unveil the power of unconscious bias: Provide comprehensive training on unconscious bias to increase awareness among team members regarding its impact on decision-making processes. By understanding their biases, individuals can better recognize them and then mitigate their influence on thoughts and actions.

b. Cultivate a culture of open communication: Foster an environment that encourages open and honest dialogue in which team members feel safe expressing their thoughts, concerns, and experiences without fear of judgment or reprisal.

c. Establish avenues for seeking support: Implement a support system that enables team members to access guidance and assistance when required. This could involve designated managers, mentors, or peers who can offer support, advice, or simply a listening ear.

d. Encourage diverse perspectives: Actively invite diverse perspectives and input from colleagues of various backgrounds when making decisions or solving problems. Encouraging inclusive collaboration enables the exploration of multiple viewpoints and fosters a more comprehensive understanding of issues.

e. Embrace a code of conduct: Develop and implement a comprehensive code of conduct that embodies inclusivity and respect for diversity. Clearly communicate that adherence to this code is expected from all team members, and establish mechanisms to address violations effectively.

As an AI hero, it is your duty and responsibility to create welcoming environments that empower team members to confidently challenge the status quo, interrupt biases, and create a safe space for seeking support.

3. Respect in Action: Establish Your Ethical Standards Team

Whether you are on a personal mission or serving an organization, surrounding yourself with a team that holds you accountable is crucial to the success of your endeavors. As an AI hero, respect in action means inviting allies and champions to join you in your cause. Allies and the power of collaboration are more potent than you can imagine. Even heroes need allies. In this section, we illuminate the significance of diverse teams and inclusive workplaces and emphasize how fostering a culture of collaboration and empowerment can enhance AI innovation while mitigating biases.

The establishment of your ethical standards team serves as a cornerstone for upholding the organization's ethical standards as outlined in the foundational inclusion strategy. Comprised of members from various departments within the organization, this team takes on the vital responsibility of reviewing policies and procedures to ensure their alignment with ethical principles. Their commitment extends to making recommendations to management on ethical matters and conducting thorough investigations and hearings regarding potential violations of ethical standards.

Beyond maintaining compliance, the ethical standards team actively promotes a culture of ethical behavior throughout the organization. They are champions of integrity, offering guidance to employees on navigating ethical dilemmas and fostering an environment in which open dialogue and principled decision-making thrive. By embodying respect in action, this team plays a pivotal role in safeguarding the organization's ethical framework and nurturing trust among employees and stakeholders.

Through their collective efforts, the ethical standards team becomes a beacon of ethical conduct, championing the organization's commitment to fairness, transparency, and accountability. They are the proof that ethical behavior is not merely a set of rules, but a living principle that guides every aspect of the organization's operations. With their guidance and expertise, the organization can navigate complex ethical challenges and cultivate a workplace in which integrity is the foundation for success.

Here are some steps you can take to establish your ethical standards team:

a. Determine the purpose and scope of the team: Begin by defining the specific issues or areas the team will address, such as the ethical use of algorithms or data privacy.

b. Identify potential members: Seek individuals with relevant expertise, experience, and diverse perspectives to form the team.

c. Establish a selection process: Develop a process for selecting team members, such as soliciting nominations or conducting interviews.

d. Define the team's powers and responsibilities: Clearly outline how the team will operate, make decisions, and report to relevant stakeholders.

e. Communicate the team's establishment and purpose: Once the team is formed, communicate its existence and purpose to relevant stakeholders, including team members and senior management.

Empower your ethical standards team to uphold the organization's ethical principles and foster a culture of integrity, transparency, and accountability. Together, you will build a strong foundation for ethical decision-making and ensure that your AI initiatives align with your values and aspirations.

4. Optimize Inclusion: Develop an Algorithmic Auditing Cycle of Inquiry

Optimizing for inclusion compels us to actively engage in activism that fosters a culture of belonging and drives positive change. The term "optimizing" refers to the process of making something as effective, efficient, or beneficial as possible. It involves maximizing the performance, quality, or value of a system, process, or outcome by identifying and implementing improvements or enhancements. In this context, for AI heroes, "optimizing" refers to actively working towards creating an environment, culture, or system that fully embraces and leverages the benefits of diversity, promotes equality, and ensures that all individuals can thrive and contribute their unique talents.

This involves ensuring that AI technologies are inclusive and accessible to all individuals, regardless of their backgrounds, characteristics, or abilities. When optimizing AI for inclusion, the focus is on identifying and mitigating biases that may exist within the data, algorithms, or decision-making processes. It involves taking steps to reduce the potential for discriminatory outcomes and ensuring that AI systems are designed to provide equal opportunities and treatment for all users. Optimizing for inclusion also involves considering the impact of AI on underrepresented groups and marginalized communities. It requires actively seeking diverse perspectives and input throughout the development and deployment stages of AI systems to ensure that they address the unique needs and concerns of different user groups. It urges us to boldly challenge systemic barriers, advocate for equal access and opportunities, and dismantle structures that perpetuate inequality.

By optimizing inclusion, we strive to create a world in which every individual can fully participate, thrive, and contribute their unique talents.

An algorithmic auditing cycle of inquiry refers to the continuous process of regularly reviewing and evaluating algorithms to ensure their accuracy, fairness, and ethical use. This rigorous process involves several key steps, including identifying the algorithms in use, understanding their purpose and intended outcomes, rigorously testing and verifying their accuracy and fairness, and promptly addressing any identified issues.

The cycle of inquiry is typically ongoing, with regular reviews and evaluations being conducted to ensure that algorithms are functioning properly and producing fair and unbiased results. The results of the cycle of inquiry should be reported to relevant collaborators, including team members and senior management, to ensure that any issues or concerns are addressed in a timely manner.

Overall, an algorithmic auditing cycle of inquiry is an important process that helps ensure the accuracy and fairness of algorithms, and promotes responsible and ethical use of these tools. By regularly reviewing and evaluating algorithms, an organization can identify and address any issues that may be causing inefficiencies or reducing the effectiveness of the algorithms in use. This can help the organization get the most value out of its algorithms. Here are some steps that an organization can take to create an algorithmic auditing discipline that helps ensure the accuracy and fairness of the algorithms being used:

a. Identify the algorithms that are being used within the organization: This is the first step in creating an algorithmic auditing discipline. It may include algorithms that are used to make decisions, as well as algorithms that are used for other purposes, such as data analysis.

b. Determine the purpose and intended outcomes of each algorithm: This can help ensure that the algorithms are being used appropriately and in line with the organization's goals.

c. Establish a process for regularly reviewing and evaluating algorithms being used: This process should include steps for testing and verifying the accuracy and fairness of the algorithms, as well as identifying and addressing any issues that are discovered.

d. Identify a team or individual who will be responsible for overseeing the algorithmic auditing process: This team or individual should have the necessary skills and expertise to effectively evaluate algorithms.

e. Communicate the results of the algorithmic auditing process: This should be done regularly to keep all relevant collaborators, including team members and senior management, informed. This can help ensure that any issues or concerns are addressed in a timely manner.

5. Audit for Bias

Bias refers to a tendency or inclination, especially one that prevents impartiality or fairness. It can refer to a specific attitude or preference that influences a person's judgment, or it can refer to a systemic problem in which certain groups are unfairly disadvantaged. Bias can be conscious or unconscious, and it can have negative consequences on both individuals and society at large. Bias can affect how people treat others, how they make decisions, and how they perceive and interpret information. It is important to recognize and address bias in order to create a more fair and just society.

One example of how we are all biased is through the use of cognitive shortcuts, also known as heuristics. These are mental shortcuts that allow us to make decisions and solve problems quickly and efficiently. While heuristics can be helpful in many situations, they can also lead to biases because they rely on incomplete or stereotypical information.

Another example of bias is confirmation bias, which is the tendency to search for, interpret, or prioritize information that confirms our preexisting beliefs, while ignoring or discounting information that contradicts them. This can lead us to become overly confident in our beliefs and make us resistant to new information or perspectives.

These are just a couple of examples as to how our brain's natural biases can influence our perceptions and decision-making. There are many other types of biases that can affect how we think and act.

Algorithmic auditing for bias is a process used to identify potential biases in AI systems. This involves analyzing data, models, and algorithms in order to identify any potential sources of bias. This process is important in order to ensure that AI systems are making decisions in an objective and fair manner, free from any sources of bias. Additionally, algorithmic auditing for bias can help to identify any potential sources of discrimination or unfairness within the system. Once potential sources of bias have been identified, they can be addressed and corrected in order to ensure that the system is making decisions in an ethical and responsible manner.

To audit for bias means to evaluate algorithms or other decision-making systems to determine whether they are producing biased or unfair results. Bias in algorithms can occur for a variety of reasons, such as the use of biased or inaccurate data, or the inclusion of biased assumptions or design choices. Auditing for bias typically involves reviewing the algorithms and the input data that is used to train and test them, as well as using statistical tests and other methods to evaluate the outputs for fairness. If any biases are identified, steps can be taken to address them, such as adjusting the algorithms or implementing additional controls to mitigate against the harmful effects of bias.

Daniel Kahneman is an Israeli-American psychologist and economist who is best known for his work on the psychology of judgment and decision-making, as well as his contributions to behavioral economics. He was awarded the Nobel Memorial Prize in Economic Sciences in 2002 for his groundbreaking work in this field. Kahneman's pioneering work examined human judgment and decision-making under uncertainty, and he is most famous for his book *Thinking Fast and Slow*. He has written numerous books and articles on the topics of judgment and decision-making, and he is widely regarded as one of the most influential psychologists of the twentieth century. Professor Kahneman's work on cognitive biases describes the use of a checklist as a somewhat effective and proven way to consider bias and minimize its effects.

May I suggest that, as leaders, we consider a checklist of biases to consider while we build a plan for auditing algorithms for bias. Checklists are important to decision-making and process optimization because they provide structure and clarity to complex tasks. By providing a step-by-step guide to decision-making and process optimization, checklists can help organizations ensure that all important factors are taken into account and that development decisions are made in a systematic way. Additionally, checklists can help to prevent bias by providing an objective framework for decision-making and by explicitly outlining the steps that need to be taken. By using a checklist, we are more likely to take an objective, impartial approach to decision-making and process optimization.

Let's go over a sampling of what this auditing process looks like. As a leader, you can use this guide to help your teams answer these questions in order to consider eliminating bias. During the auditing process, be sure to check for the following twelve types of bias:

Self-interested Biases

Self-interested bias, also known as self-serving bias, is the tendency to attribute our successes to our own abilities and traits, while attributing our failures to external factors. This bias can lead us to overestimate our own abilities and accomplishments, and it can also make us more resistant to criticism or feedback. Self-interested bias can affect our perceptions of ourselves and our abilities, as well as our relationships with others. It can also influence our decision-making, as we may be more likely to make choices that benefit us, even if they are not in the best interest of others. Self-interested bias can be a natural human tendency, and it is important to be aware of it and try to mitigate its effects in order to make more objective and fair decisions. Ask questions similar to these:

- Are we attributing our successes to our own abilities and traits, while attributing our failures to external factors?

- Are we overestimating our own abilities or accomplishments?

- Are we resistant to criticism or feedback, or do we see it as a personal attack?

- Are we considering the potential impact of our actions on others, or are we primarily focused on our own interests?

Asking these questions can help the team become aware of any self-serving bias they may have and help you make more objective and fair decisions.

Affect Heuristic

The affect heuristic is a mental shortcut that refers to the influence of our emotions on our judgments and decision-making. It occurs when we allow our feelings about a particular event or idea to affect our perception of it, rather than relying on more objective or logical factors.

For example, if we have a positive emotional reaction to a product or brand, we may be more likely to perceive it as of higher quality or value, even if this is not necessarily the case. Similarly, if we have a negative emotional reaction to something, we may be more likely to view it negatively, regardless of its actual merits.

The affect heuristic can lead us to make biased decisions that are not based on rational or objective considerations. It is important to be aware of this bias and try to mitigate its influence by considering multiple perspectives and basing decisions on evidence and logical reasoning. Ask questions similar to the following:

- Has the team fallen in love with its proposal?

- Are our emotions influencing our perceptions and decision-making in this situation?

- Are we considering multiple perspectives, or are we primarily focused on our own emotional reactions?

- Are we basing our decisions on evidence and logical reasoning, or are we allowing our emotions to drive our choices?

- Are we being objective and unbiased in our evaluation of the situation?

These questions are important because the team could be making decisions that are heavily influenced by their current emotions. Emotions are influential in all types of decision-making, especially when it comes to technology development. Asking these questions can help a team become aware of the influence of affective biases, guard against them, and make more objective and fair decisions. It can also be helpful to seek feedback from others and to consider multiple perspectives in order to reduce the impact of the affect heuristic.

Groupthink

Groupthink is a phenomenon that occurs when group members prioritize group harmony and conformity over critical thinking and decision-making. In a group suffering from groupthink, members may be reluctant to express their own opinions or concerns, or they may go along with the majority opinion even if they disagree with it. This can lead to poor decision-making and a lack of creativity within the group.

Groupthink can be caused by a variety of factors, including a desire to avoid conflict, a need to maintain group cohesion, or a lack of diversity within the group. It can be particularly dangerous in situations in which the stakes are high, as it can lead to faulty decisions that can have serious consequences.

To prevent groupthink, it is important for group members to feel free to express their own opinions and for there to be open and honest communication within the group. It is also helpful to encourage diversity of thought and to have a clear process for decision-making that allows for the consideration of multiple perspectives.

Here are a few questions a team can ask themselves to check for groupthink:

- Are all team members comfortable expressing their own opinions and ideas?

- Is there open and honest communication within the group, or are certain opinions or perspectives being dismissed or suppressed?

- Is the group considering multiple perspectives and options, or are they only focusing on a single course of action?

- Is the group being too influenced by a desire for harmony or conformity, and not considering the potential risks or drawbacks of their decisions?

- Were there differing opinions within the team before a final decision was made? Were they explored adequately?

This is important because groupthink, even in minor cases, could trigger decisions that are not ideal and that may ignore critical information.

Saliency Bias

Is the decision being overly influenced by an analogy or a story of a memorable success? Saliency bias is the tendency to give more attention to information or stimuli that stand out or are more noticeable. This can occur because our brain is naturally drawn to information that is unusual or unexpected, or because we are more likely to remember things that are more noticeable or salient.

Saliency bias can affect our decision-making and our perception of the world around us, as we may give more weight to information that is more noticeable, even if it is not necessarily the most important or relevant. For example, we may be more likely to remember a particularly unusual or striking advertisement, even if it is not for a product that we need or want.

To mitigate the effects of saliency bias, it is important to try to consider all relevant information, rather than just focusing on the most noticeable or striking aspects of a situation. It can also be helpful to seek feedback from others and to consider multiple perspectives in order to get a more complete and objective view of a situation.

Here are a few questions a team can ask themselves to check for saliency bias:

- Are we giving too much weight to information or stimuli that stand out or are more noticeable, even if they are not necessarily the most important or relevant?

- Are we considering all relevant information, or are we only focusing on a few standout details?

- Are we considering multiple perspectives, or are we only considering information that confirms our preexisting beliefs or assumptions?

- Are we being objective and unbiased in our evaluation of the situation, or are we allowing our own biases or emotions to influence our judgment?

Asking these questions can help a team become aware of any saliency bias, mitigate against it, and make more objective and fair decisions. It can also be helpful to seek feedback from others and to consider multiple perspectives in order to reduce the impact of saliency bias.

Confirmation Bias

Confirmation bias is the tendency to search for, interpret, or prioritize information that confirms our preexisting beliefs, while ignoring or discounting information that contradicts them. This bias can lead us to become overly confident in our beliefs and make us resistant to new information or perspectives.

Confirmation bias can affect our decision-making and our perceptions of the world around us, as we may selectively seek out or pay more attention to information that confirms our preexisting beliefs, rather than considering a more complete or objective view of a situation.

This can lead us to make biased or flawed decisions, as we are not considering all relevant information.

To mitigate against confirmation bias, it is important to try to consider multiple perspectives and to be open to new information that may challenge our preexisting beliefs. It can also be helpful to seek feedback from others and to be aware of our own biases in order to get a more complete and objective view of a situation.

Ask the team questions similar to:

- Are credible alternatives included along with the recommendation?

- Are we considering multiple perspectives and being open to new information that may challenge our preexisting beliefs?

- Are we seeking out information that confirms our preexisting beliefs, or are we also considering information that contradicts them?

- Are we considering all relevant information, or are we only focusing on information that confirms our preexisting beliefs?

- Are we being objective and unbiased in our evaluation of the situation, or are we allowing our own biases to influence our judgment?

Asking these questions can help a team become aware of any confirmation bias, mitigate against it, and make more objective and fair decisions. It can also be helpful to seek feedback from others and to consider multiple perspectives in order to reduce the impact of confirmation bias.

Availability Bias

Availability bias is the tendency to estimate the likelihood of an event based on how easily we can recall similar events from memory. This bias can lead us to overestimate the likelihood of events that are more easily remembered or that we have heard more about, while underestimating the likelihood of events that are less familiar or that we have heard less about. Availability bias can affect our perceptions and decision-making, as we may base our judgments on information that is more readily available to us, rather than considering a more complete or objective view of the situation. This can lead to biased or flawed conclusions.

To mitigate against availability bias, it is important to consider all relevant information and to be aware of the potential influence of our own past experiences and exposures on our perceptions and judgments. It can also be helpful to seek feedback from others and to consider multiple perspectives in order to get a more complete and objective view of a situation.

Here are a few questions a team can ask themselves to check for availability bias:

- Are we basing our judgments on information that is readily available to us, or are we considering all relevant information?

- If you had to make this decision again in a year's time, what additional information would you want, and can you get it now?

- Are we aware of the potential influence our own past experiences and exposures have on our current perceptions and judgments?

- Are we considering multiple perspectives, or are we only considering information that confirms our preexisting beliefs or assumptions?

- Are we being objective and unbiased in our evaluation of the situation, or are we allowing our own biases to influence our judgment?

Asking these questions can help a team become aware of any availability bias, mitigate against it, and make more objective and fair decisions. It can also be helpful to seek feedback from others and to consider multiple perspectives in order to reduce the impact of availability bias.

Anchoring Bias

Anchoring bias is the tendency to rely too heavily on the first piece of information we receive when making a decision. This bias occurs because our brains naturally look for shortcuts to help us make decisions quickly and efficiently, and we tend to anchor our judgments to the first piece of information we receive. This can lead us to give too much weight to that initial piece of information, even if it is not necessarily the most important or relevant. Anchoring bias can affect our decision-making and our perceptions of the world around us, as we may be more influenced by the first piece of information we receive, rather than considering a more complete or objective view of a situation. This can lead to biased or flawed conclusions.

To mitigate against anchoring bias, it is important to be aware of it and to consider all relevant information when making decisions. It can also be helpful to seek feedback from others and to consider multiple perspectives in order to get a more complete and objective view of a situation.

Here are a few questions a team can ask themselves to check for anchoring bias:

- Are we giving too much weight to the first piece of information we received, or are we considering all relevant information?

- Are we considering multiple perspectives, or are we only focusing on a single piece of information?

- Are we being objective and unbiased in our evaluation of the situation, or are we allowing our own biases to influence our judgment?

- Are we open to new information and willing to revise our conclusions, if necessary, or are we anchored to our initial beliefs?

- Do we know where the data came from?

- Can there be unsubstantiated numbers or any extrapolation of the data from history? If this is the case, then re-anchor the figures and request new analyses.

Asking these questions can help a team become aware of any anchoring bias, mitigate against it, and make more objective and fair decisions. It can also be helpful to seek feedback from others and to consider multiple perspectives in order to reduce the impact of anchoring bias.

Halo Effect

The halo effect is a cognitive bias that occurs when our overall impression of someone or something influences how we perceive that person or thing in specific domains. For example, if we have a positive overall impression of someone, we may view them as more intelligent, competent, or attractive, even if we have no evidence to support these specific judgments. Similarly, if we have a negative overall impression of someone, we may view them as less intelligent, competent, or attractive.

The halo effect can influence our perceptions and judgments of people, products, and organizations, thereby affecting our decisions and behaviors. It is important to be aware of the halo effect and to try to consider specific attributes or characteristics separately, rather than being influenced by our overall impression of someone or something.

Here are a few questions a team can ask themselves to check for the halo effect:

- Are we allowing our overall impression of someone or something to influence how we perceive that person or thing in specific domains?

- Are we considering specific attributes or characteristics separately, or are we being influenced by our overall impression of someone or something?

- Are we being objective and unbiased in our evaluation of the situation, or are we allowing our own biases to influence our judgment?

- Are we considering multiple perspectives and multiple sources of information, or are we relying on a single source or perspective?

Asking these questions can help a team become aware of any halo effect, mitigate against it, and make more objective and fair decisions. It can also be helpful to seek feedback from others and to consider multiple perspectives in order to reduce the impact of the halo effect.

Sunk-Cost Fallacy and the Endowment Effect

The sunk-cost fallacy, also known as the sunk-cost bias or sunk-cost effect, refers to the tendency to continue investing time, money, or other resources into a project or decision because we have already invested a significant amount of resources in it, even if it is no longer in our best interests to do so. This bias occurs because we tend to feel that we have to justify our past decisions by continuing to invest in them, even if the situation has changed.

The endowment effect, on the other hand, is a cognitive bias that refers to the tendency to overvalue something simply because we own it or are invested in it. This bias occurs because we tend to place a higher value on things that we possess or are connected to, even if they have no objective value.

Both the sunk-cost fallacy and the endowment effect can influence our decision-making and lead us to make choices that are not in our best interests. It is important to be aware of these biases, mitigate against them, and consider all relevant information and options when making decisions, rather than being influenced by past investments or our attachment to certain outcomes.

Here are a few questions a team can ask themselves to check for the sunk-cost fallacy:

- Are we continuing to invest time, money, or other resources in a project or decision because we have already invested a significant amount in it, even if it is no longer in our best interests to do so?

- Are we considering the current situation and all relevant information, or are we being influenced by our past investments or decisions?

- Are we being objective and unbiased in our evaluation of the situation, or are we allowing our own biases to influence our judgment?

- Are we open to alternative options and willing to consider new information, or are we committed to a specific course of action because of our past investments?

Asking these questions can help a team become aware of any sunk-cost fallacy, mitigate against it, and make more objective and fair decisions. It can also be helpful to seek feedback from others and to consider multiple perspectives in order to reduce the impact of the sunk-cost fallacy.

Optimistic Biases and Competitor Neglect

Optimistic bias is the tendency to overestimate the likelihood of positive outcomes, and underestimate the likelihood of negative outcomes. This bias can lead us to be overly optimistic about the future and to underestimate the risks or challenges that we may face.

Competitor neglect is the tendency to underestimate the abilities or resources of our competitors, or to underestimate the potential impact that they may have on our own plans or goals. This bias can lead us to underestimate the level of competition that we face and to be overly confident in our own abilities or resources.

Both optimistic bias and competitor neglect can affect our decision-making and our perceptions of the world around us, and they can lead us to make flawed or unrealistic judgments. It is important to be aware of these biases, mitigate against them, and try to consider all relevant information and perspectives when making decisions, rather than being influenced by overly optimistic or overly confident beliefs.

Here are a few questions a team can ask themselves to check for optimistic bias and competitor neglect:

- Are we overestimating the likelihood of positive outcomes and underestimating the likelihood of negative outcomes?

- Are we underestimating the abilities or resources of our competitors, or the potential impact that they may have on our plans or goals?

- Are we being objective and unbiased in our evaluation of the situation, or are we allowing our own biases to influence our judgment?

- Are we considering all relevant information and perspectives, or are we only focusing on a single perspective or set of assumptions?

Asking these questions can help a team become aware of any optimistic bias or competitor neglect, mitigate against them, and make more realistic and fair decisions. It can also be helpful to seek feedback from others and to consider multiple perspectives in order to reduce the impact of these biases.

Disaster Neglect

Disaster neglect is a cognitive bias that refers to the tendency to underestimate the likelihood or consequences of potential disasters or negative events. This bias can lead us to be overly complacent and unprepared for disasters or other negative events, as we did not adequately consider the potential risks. This can have serious consequences; because we are unprepared for disasters or other negative events, our effective planning or response is either impaired or nonexistent.

It is important to be aware of disaster neglect and to consider all relevant information and perspectives when evaluating potential risks and preparing for negative events. It is also important to be proactive in identifying and mitigating the potential risks, rather than assuming that disasters or negative events will not occur.

Here are a few questions a team can ask themselves to check for disaster neglect:

- Are we underestimating the likelihood or consequences of potential disasters or negative events?

- Are we being proactive in identifying and mitigating potential risks, or are we assuming that disasters or negative events will not occur?

- Are we considering all relevant information and perspectives when evaluating potential risks and preparing for negative events?

- Are we being objective and unbiased in our evaluation of the situation, or are we allowing our own biases to influence our judgment?

Asking these questions can help a team become aware of any disaster neglect, mitigate against it, and make more realistic and prepared decisions. It can also be helpful to seek feedback from others and to consider multiple perspectives in order to reduce the impact of this bias.

Loss Aversion

Loss aversion is a cognitive bias that refers to the tendency to strongly prefer avoiding losses and acquiring equivalent gains. This bias occurs because we tend to feel the pain of loss more strongly than the pleasure of gain, and we often prioritize avoiding losses over seeking out opportunities for gain.

Loss aversion can affect our decision-making and our perceptions of risk, as we may be more hesitant to take risks or to pursue opportunities if there is a potential for loss. This bias can also influence our behaviors, as we may be more likely to stick with the status quo or to avoid change if it involves the potential for loss.

It is important to be aware of loss aversion, to mitigate against it, and to try to consider all relevant information and options when making decisions, rather than being overly influenced by a fear of loss. It can also be helpful to consider the potential long-term benefits and consequences of a decision, rather than focusing solely on the immediate costs or potential losses.

Here are a few questions a team can ask themselves to check for loss aversion:

- Are we prioritizing avoiding losses over seeking out opportunities for gain?

- Are we being overly influenced by a fear of loss or a desire to maintain the status quo, rather than considering all relevant options and information?

- Are we considering the potential long-term benefits and consequences of a decision, or are we only focusing on the immediate costs or potential losses?

- Are we being objective and unbiased in our evaluation of the situation, or are we allowing our own biases to influence our judgment?

Asking these questions can help a team become aware of any loss aversion, mitigate against it, and make more balanced and fair decisions.

Consider the Dimensions of Diversity

For all these biases, it can also be helpful to seek feedback from others outside of the ethical standards team and to consider multiple perspectives in order to reduce the impact of bias. This is important because the dimensions of diversity are vast; it encompasses many different aspects of an individual's identity, experiences, and background. By considering perspectives from all dimensions of diversity, we can get a more complete and nuanced understanding of a situation, and we can make more inclusive and fair decisions.

Consider how the development would impact groups based on characteristics such as race, ethnicity, sex, gender, sexual orientation, disability, mental health, history of substance abuse, body size, immigration status, nationality, language, culture, poverty, socioeconomic status, age, religion, spirituality, faith traditions, prison record, rural residents, interprofessional communication, to name a few.

In addition, by considering perspectives from all dimensions of diversity, we can foster a more inclusive and respectful environment, and we can recognize and value the unique contributions and experiences of all individuals. This can lead to a more cohesive and productive team, and it can help to promote a sense of belonging and engagement among team members.

Considering perspectives from all dimensions of diversity is also important because it helps to challenge our own biases and assumptions, and it can help us to develop a more comprehensive and empathetic understanding of the world around us.

Overall, auditing for bias is an important step in ensuring that algorithms and other decision-making systems are fair and unbiased, and that they do not discriminate against certain groups or individuals.

6. Audit for Transparency

In the context of organizations and institutions, transparency means being accountable and open to public scrutiny, allowing others to see how decisions are being made and resources are being used. Transparency is important because it helps to build trust and accountability, and it allows for the fair and equitable treatment of individuals and groups. It can also help to promote fairness and prevent corruption or abuse of power. In order to promote transparency, organizations and institutions may make information and processes available to the public, and may be subject to oversight or regulation to ensure that they are operating in a transparent and accountable manner.

Algorithmic auditing for transparency is the process of inspecting algorithms and their associated data sources for opacity, to ensure that they are free from bias and are making decisions fairly and accurately. Opaque algorithms are harmful because they are not transparent and, thus, can be used to perpetuate bias, discrimination, and unfairness. Opaque algorithms can be used to reinforce existing inequalities and bias, as they are not transparent; therefore, their decision-making processes and results are difficult to understand and challenge. Opaque algorithms can also lead to incorrect results, as they are not transparent and, thus, can be vulnerable to errors. Furthermore, opaque algorithms can be used to manipulate outcomes and lead to decisions that are not in the best interest of the people they are supposed to serve.

Algorithmic auditing for transparency involves testing algorithms to see if they are meeting their goals and objectives, while also checking to make sure they are not exhibiting any discriminatory or biased behavior. Auditors examine algorithms to look for potential issues, such as overfitting, underfitting, or other errors that could lead to incorrect or misleading results.

Similar to algorithmic bias, we can audit algorithms for transparency by following a checklist. This checklist should include steps such as analyzing the data used to train the algorithm, testing the algorithm for bias and fairness, assessing the algorithm to see if it is explainable, and testing for accuracy.

Here are is a checklist that can be used to audit algorithms for transparency:

Transparency of Architecture

This is a measure of how well the structure of the algorithm is known (or knowable) to collaborators, including inputs and outputs. The questions below can be used as a starting point when evaluating the transparency of an algorithm's architecture. It may also be helpful to consult with the developers or other experts to review any technical documentation that is available. The goal of checking for transparency is to understand how the algorithm works and to ensure that it is being used appropriately and ethically. Questions to ask include, but are not limited to:

- How does the algorithm make decisions?

- What data is used to train and test the algorithm?

- What are the details of the equation used to develop the algorithm?

- Is the data representative and free of bias?

- How is the algorithm validated and tested for accuracy and fairness?

- Are the results of the algorithm transparent and explainable?

- How are the results of the algorithm used and incorporated into decision-making processes?

- Are there any safeguards in place to mitigate the risk of this algorithm?

Explainable and Interpretable

"Explainable" refers to the degree to which the outputs and decision-making processes of an algorithm can be easily understood and explained to a general audience. This includes the ability to understand how the algorithm makes decisions and what factors it considers when making those decisions. "Interpretable," on the other hand, refers to the ability to understand the meaning and significance of the outputs of an algorithm. This includes the ability to understand how the outputs relate to the input data and how they can be used to inform decision-making.

These two terms are related but distinct concepts. "Explainable" focuses on the inner workings of the algorithm, while "interpretable" focuses on the meaning and significance of the outputs. Both are important for ensuring that algorithms are transparent. For any given use of the algorithm, both reflect how well one can know why and how a particular output was given. The questions below are a sampling of those that can be used when evaluating how explainable and interpretable an algorithm is and if its outputs can be easily understood and interpreted by significant collaborators.

- How and by whom was the algorithm developed?

- Are the developers clearly identifiable, technically prepared, and qualified for this system development?

- Is there a clear process for designing and testing the algorithm, or is it a black box that is not well understood?

- Can the results of the algorithm be easily understood and explained to someone who is not an expert in the field?

- How are the results of the algorithm presented?

- How are the results used and incorporated into the decision-making process?

- Are there any safeguards in place to mitigate the risk in this algorithm?

Transparency of Use

Algorithmic transparency of use is an important concept in the field of AI. It refers to making sure that users are aware of when algorithms are being used to make decisions that impact their lives. Transparency of use is important because it gives users the ability to understand how and why decisions are being made, and allows them to hold algorithms accountable for any mistakes or biases that may arise. Examples include online ad targeting or NSA (National Security Agency) surveillance—which until recently was opaque to most users—versus credit scoring, which is widely known. Additionally, algorithmic transparency of use helps to ensure that algorithms are making decisions based on accurate and up-to-date data, and that these decisions are not influenced by any biased or discriminatory factors.

Questions to include in the checklist include, but are not limited to:

- How transparent is the fact that the algorithm is being used?

- Is the use of the algorithm clearly communicated to users, or is it hidden from them?

- Has a privacy statement been developed that outlines how user data is collected, used, and protected when the algorithm is being used?

- Are the inputs, outputs, and decision-making processes of the algorithm clearly defined and documented?

- Is the algorithm's performance and accuracy publicly available or disclosed?

- Does the algorithm have any potential biases or discriminatory effects that could affect certain groups of people unfairly? If so, how is this being addressed?

- Is there a process in place for users to provide feedback or report problems with the algorithm?

- Are there any safeguards in place to mitigate the risk in this algorithm?

Transparency of Data Use and Collection

In the context of data use and collection, algorithmic transparency is concerned with how data is collected, used, and protected by an algorithm. Algorithmic transparency of data use and collection is the process of making sure that users are aware of how algorithms are obtaining and using their data. It is important for users to be able to understand the process that algorithms are using to collect and analyze data in order to ensure that their data is being used ethically and responsibly. Additionally, this transparency can help to prevent algorithms from using data in ways that are discriminatory or biased. This includes issues with the types of data being collected, how it is analyzed, and how the results of the analysis are used.

Important questions to include in the checklist include, but are not limited to:

- Can the users see what types of data are being collected and how it is being used?

- Do the users know how long the data is being stored, and for what purpose?

- Are the users able to determine how their data is used?

- What measures have been put in place to ensure the security and integrity of the algorithm and the data it processes?

- How is the data collected and stored, and by whom?

- Is the data collected fairly and ethically, with appropriate consent and protections in place?

- Is the data collection and use clearly communicated to users?

- Are users given the opportunity to review and correct their data, or to request that it be deleted or removed from the system?

- Are users aware of any third parties that may have access to their data, and for what purposes?

- Have all collaborators considered what further processing or inferences will be made using the data, and for what purpose?

- Is the data retention policy clearly communicated to users?

- Are users given the opportunity to opt out of data collection or usage, or to set privacy preferences?

- Is there a process in place for handling data breaches or other security incidents?

- How is the use of the data governed and regulated?

- Is there a clear process for oversight and accountability?

Good communication and democratic rule can help ensure a society that is self-correcting. When data is transparent and easily accessible to the public, individuals and organizations can use it to hold decision-makers accountable and to identify and address any issues or concerns.

For example, if an algorithm is being used to make decisions that impact individuals or groups, it is important that the inputs, outputs, and decision-making processes of the algorithm are transparent and easily understood. This allows collaborators to evaluate the algorithm for fairness and to identify and address any biases or issues that may be present. By auditing algorithms for transparency, organizations can ensure that their algorithms are functioning correctly and are transparent to ensure that they are being used appropriately and ethically.

7. Audit for Loss of Privacy

"Loss of privacy" refers to the situation in which an individual's personal information or activities are made known to others without their consent. This can occur through the unauthorized disclosure of personal information, the tracking of an individual's activities or location, or the monitoring of their communications.

A loss of privacy can have serious consequences for individuals, including a loss of control over their personal information, the potential for identity theft or fraud, and a feeling of vulnerability or lack of security. It can also have broader social and economic impacts, as the loss of privacy can erode trust and undermine the foundations of a free and open society.

To protect privacy, it is important for individuals to be aware of how their personal information is being collected and used, and to take steps to protect it. It is also important for organizations and institutions to respect the privacy of individuals and to handle personal information responsibly.

The link between algorithms and the loss of privacy is a growing concern in today's digital world. With the rise of big data, more and more organizations are collecting personal data and using algorithms to process it. This can lead to loss of privacy, as organizations can use algorithms to identify patterns, predict behavior, and target their services to individuals based on their data. Furthermore, algorithms can be used to create profiles of people that can be used to make decisions about them, such as whether or not to grant them a loan or offer them a job. As algorithms become more powerful and ubiquitous, it is becoming increasingly important to ensure that privacy is protected, and that algorithms are not used to perpetuate bias or unfairness.

Organizations can audit algorithms for loss of privacy by following a checklist of best practices. This checklist should include steps such as ensuring that the data used to train the algorithm is redacted to protect the privacy of individuals. Consider the following:

Potential for Abuse

The potential for abuse of this data rises when it is used in ways that are not transparent or are not in the best interests of the individuals whose data is being collected and analyzed. There are several ways in which algorithmic loss of privacy can be abused. For example, the data collected by algorithms could be used to discriminate against certain individuals or groups, either intentionally or unintentionally. It could also be used to manipulate or influence individuals in ways that are not transparent or that they might not fully understand. Additionally, the data collected by algorithms could be misused by hackers or other malicious actors, who could access and exploit it for their own purposes.

Organizational development teams can ask the following questions to check for the potential abuse of algorithmic systems:

- Does the algorithm disproportionately affect certain groups of people?

- Does the algorithm have the potential to perpetuate or amplify existing biases?

- How transparent is the decision-making process of the algorithm?

- How can we ensure that the algorithm is accountable and explainable?

- How can we test the algorithm for bias and fairness?

- What is the process for addressing and mitigating any issues that are identified with the algorithm?

- How will the use of the algorithm be monitored to ensure that it is not being abused?

It is important for organizations to carefully consider the potential consequences of using algorithmic systems and to put safeguards in place to prevent abuse. It is also important for users to be aware of the potential risks associated with algorithmic loss of privacy and to take steps to protect their own data and privacy. This may include being mindful of the types of data they share online, using privacy-enhancing technologies, and advocating for stronger privacy protections at the individual and policy level. Forward-thinking organizations may consider offering this information to users.

Infringement of (Legal) Rights

Algorithmic infringement of legal rights refers to the way that algorithms and other automated systems can potentially violate an individual's legal rights. This can occur in a variety of ways, depending on the specific context and the rights in question.

One way in which algorithmic infringement of legal rights can occur is through the use of biased or discriminatory algorithms. For example, an algorithm used in the criminal justice system that is based on biased data or that is not properly calibrated could result in individuals being unfairly treated or punished. Similarly, an algorithm used in hiring or other types of decision-making that is biased against certain groups of people could result in those individuals being denied opportunities or benefits that are rightfully theirs.

Another way in which algorithmic infringement of legal rights can occur is through the misuse or abuse of personal data collected by algorithms. For example, if an algorithm is collecting and analyzing data about an individual without their knowledge or consent, this could violate their rights to privacy and informational autonomy. Additionally, if the data collected by an algorithm is used to make decisions that impact an individual's life, such as decisions related to their employment or creditworthiness, there is a risk that their legal rights could be infringed upon if the algorithm is not fair or transparent.

Overall, it is important to ensure that algorithms and other automated systems do not infringe upon the legal rights of individuals, and that steps are taken to prevent and address any such infringement that may occur. This may involve efforts to ensure that algorithms are unbiased and transparent, as well as efforts to protect the privacy and data rights of individuals.

There are a number of questions that development teams can ask to check for algorithmic infringement of legal rights:

- Does the algorithm discriminate against certain groups of people on the basis of protected characteristics such as race, gender, age, or disability?

- Does the algorithm infringe on individuals' privacy rights by collecting or using personal data in an unauthorized or inappropriate manner?

- Does the algorithm violate intellectual-property rights by using protected content without permission?

- Does the algorithm comply with all relevant laws and regulations, such as those related to data protection, consumer protection, and antitrust?

- Does the algorithm have the potential to cause harm to individuals or society, and if so, have steps been taken to mitigate these risks?

- Have the appropriate permissions and approvals been obtained for the use of the algorithm, including any necessary consent from individuals whose data is being processed?

- Have appropriate measures been put in place to ensure the security and integrity of the data being processed by the algorithm?

By asking these questions, development teams can identify potential issues that may need to be addressed in order to ensure that the algorithm does not infringe on legal rights.

Security of Access to the Data

Algorithmic security of access refers to the ability of an algorithm to protect the privacy of the data it processes. In other words, it is concerned with ensuring that unauthorized parties do not have access to the data and that the data is not used in a way that violates the privacy of the individuals it concerns. This can be achieved through a variety of means, including the use of encryption, secure protocols for transmitting and storing data, and access controls that limit who has permission to view or use the data.

One important aspect of algorithmic security of access is ensuring that the algorithm itself is secure and cannot be easily exploited by attackers. This can involve measures, such as code review and testing, to identify and fix vulnerabilities, as well as using secure coding practices to minimize the risk of introducing vulnerabilities into the algorithm.

Another important aspect is ensuring that the data is only accessed and used by parties who are authorized to do so. This can involve the use of access controls and permissions systems to limit who has access to the data, as well as the implementation of policies and procedures to ensure that the data is only used for authorized purposes.

Here are some questions that development teams can ask to check for algorithmic security of access to the data:

- Who has access to the data being processed by the algorithm, and what level of access do they have?

- Are there appropriate access controls in place to limit who can access the data and what they can do with it?

- Is access to the data logged and monitored, and are there mechanisms in place to detect and prevent unauthorized access?

- Are there policies and procedures in place to ensure that the data is only accessed and used for authorized purposes?

- Is the data being transmitted and stored securely, and are appropriate measures in place to protect it from unauthorized access or tampering?

- Have appropriate permissions and approvals been obtained for the use of the data, including any necessary consent from individuals whose data is being processed?

By asking these questions, development teams can ensure that the algorithm is being used in a way that is secure and compliant with relevant laws and regulations.

Security of Access to Usage

Algorithmic security of access refers to measures that are taken to ensure that only authorized parties are able to use the algorithm and access the associated data. This can involve the use of authentication mechanisms to verify the identity of users, as well as the implementation of access controls and permissions systems that limit which users are able to access specific resources.

There are several ways in which algorithmic security of access can be achieved. One common approach is to use passwords or other forms of authentication to verify the identity of users before allowing them to access the algorithm or data. This can be combined with other measures such as two-factor authentication, which requires users to provide additional evidence of their identity, beyond just a password.

Another approach is to use access controls and permissions systems to limit which users are able to access specific resources. These systems typically involve assigning users to different groups or roles, with each group or role having different levels of access to the algorithm and data. For example, a system might have a group of administrators who have full access to all resources, while other users might have more limited access, such as the ability to view but not modify certain data.

Overall, the goal of algorithmic security of access is to ensure that only authorized parties are able to use the algorithm and access the associated data, and that these parties are only able to access the resources that they are authorized to access.

Here are some questions that development teams can ask to check for algorithmic security of access:

- Who is able to use the algorithm, and what level of access do they have?

- Can people outside the organization use it, and with what restrictions?

- Are there appropriate authentication mechanisms in place to verify the identity of users and ensure that only authorized parties are able to use the algorithm?

- Are there access controls and permissions systems in place to limit which users are able to access specific resources or perform certain actions?

- Is access to the algorithm logged and monitored, and are there mechanisms in place to detect and prevent unauthorized access or misuse?

- Are there policies and procedures in place to ensure that the algorithm is only used for authorized purposes?

- Is the algorithm being used in a way that is consistent with relevant laws and regulations, such as those related to data protection and consumer protection?

By asking these questions, development teams can ensure that the algorithm is being used in a secure and compliant manner.

Security of Data Outputs and Inferences

Algorithmic security of data inferences refers to measures that are taken to ensure that the data inferences produced by an algorithm are accurate and reliable. This can be particularly important when the inferences are used to make decisions that have significant consequences, such as in the criminal justice system or in the hiring process.

There are several ways in which algorithmic security of data inferences can be achieved. One approach is to use data validation and cleansing techniques to ensure that the input data is accurate and free of errors. This can involve checking the data for consistency, completeness, and accuracy, as well as correcting any errors or inconsistencies that are found.

Another approach is to use algorithms that are designed to be robust and resistant to errors or biases. For example, an algorithm might be designed to handle missing or incomplete data in a way that does not produce misleading inferences.

It is also important to regularly test and evaluate the accuracy of the inferences produced by the algorithm, and to take corrective action if any issues are identified. This can involve conducting independent audits or evaluations, as well as incorporating feedback from users of the algorithm.

Overall, the goal of algorithmic security of data inferences is to ensure that the inferences produced by an algorithm are accurate and reliable, and that they can be trusted to support important decision-making processes.

Here are some questions that development teams can ask to check for algorithmic security of data outputs and inferences:

- Are the data inputs to the algorithm accurate and free of errors?

- Is the algorithm designed to be robust and resistant to errors or biases?

- Have appropriate measures been taken to validate and cleanse the data inputs to the algorithm?

- Is the algorithm being used in a way that is consistent with relevant laws and regulations, such as those related to data and consumer protection?

- Have steps been taken to ensure that the algorithm is not used to make decisions that have significant consequences (for example, in the criminal justice system or in the hiring process) without being tested and validated for accuracy?

- Have independent audits or evaluations been conducted to verify the accuracy of the data outputs and inferences produced by the algorithm?

- Have measures been put in place to correct any errors or biases that are identified in the data outputs or inferences produced by the algorithm?

By following these checklists for loss of privacy auditing, developers can ensure that their algorithms are secure and accurate, and that privacy is not compromised.

8. Audit for Dependency

"Dependency" refers to the state of relying on someone or something for support or assistance. It can involve relying on others for emotional or psychological support, or it can involve relying on external resources or systems for basic needs or services.

Dependency can be a natural and normal part of human relationships and societies, as people often rely on others for various forms of support and assistance. However, excessive or unhealthy dependency can be a problem, as it can limit an individual's independence and autonomy, and it can lead to an unhealthy reliance on others. It is important to find a healthy balance between independence and reliance on others and systems. It can also be helpful to cultivate a sense of self-reliance and to develop the skills and resources needed to meet one's own needs and goals.

Algorithms can be used to automate certain processes, allowing people to focus on more important tasks. However, algorithms can also lead to people becoming overly reliant on them and reducing their critical-thinking skills. Furthermore, algorithms can be used to make decisions that may not be in the best interest of the people they are supposed to serve. This can lead to people becoming too reliant on algorithms and losing their ability to make decisions for themselves.

Here are some questions that development teams can ask to check for algorithmic auditing for dependency. By asking these questions, development teams can ensure that the use of the algorithm is appropriate and does not lead to dependency or reliance on the algorithm to the exclusion of other approaches.

- Is the algorithm being used to make decisions that have significant consequences for individuals or organizations?

- Are there alternative approaches or methods for making these decisions that do not rely on the algorithm?

- Have the potential risks and consequences of relying on the algorithm been thoroughly evaluated and addressed?

- Is there a plan in place for addressing any issues or failures that may arise with the algorithm, and is there a way to switch to alternative approaches if necessary?

- Have measures been put in place to ensure that the algorithm is transparent and explainable so that users can understand how it is making decisions and why?

- Are there safeguards in place to prevent individuals or organizations from becoming overly reliant on the algorithm, such as by encouraging the use of human judgment and decision-making in conjunction with the algorithm?

Promote Critical Thinking for Organizational Success

In addition to the algorithmic audit for dependency, organizations should also ensure that the algorithm is designed to encourage people to use their own judgment and critical-thinking skills. Critical thinking is important for organizational success because it helps individuals and teams make well-informed, effective decisions. When people are able to think critically, they are able to analyze and evaluate information and arguments, identify and assess risks and benefits, and consider multiple perspectives on a problem or issue. This enables them to make better decisions that are based on evidence and reason, rather than on gut instincts or biases.

As a leader, you can work with the ethical standards team to develop a process of promoting critical thinking within your team to increase decision-making effectiveness, promote innovation, and adapt to change effectively. Here is a step-by-step process to consider:

- Establish clear guidelines for the use of algorithms:

  The organization should establish clear guidelines for when and how algorithms should be used, as well as how their outputs should be validated and incorporated into decision-making processes.

- Educate employees on the limitations of algorithms:

  Employees should be made aware of the limitations of algorithms, including the potential for biases and errors.

  This can help prevent overreliance on algorithmic outputs.

- Encourage diverse perspectives:

    Encourage diverse perspectives and input from a wide range of individuals when making decisions that involve algorithms.

    This can help ensure that a variety of viewpoints are considered and can help mitigate the risk of bias in algorithmic outputs.

- Monitor and audit algorithms:

    Implement processes to regularly monitor and audit algorithms to ensure that they are functioning properly and producing fair and unbiased results.

- Foster a culture of critical thinking:

    Encourage employees to think critically about the outputs of algorithms and to consider the potential consequences of relying on them.

    Encourage people to evaluate and analyze information and ideas, rather than simply accepting them at face value.

- Encourage curiosity and questions:

Encourage people to ask questions and seek out answers for themselves, rather than simply accepting what they are told.

- Encourage risk-taking:

  Encourage people to take risks and try new things, even if they are not certain of the outcome.

- Encourage creativity:

  Encourage people to think creatively and come up with original ideas.

- Encourage collaboration:

  Encourage people to work with others and consider different perspectives.

- Encourage self-reflection:

  Encourage people to reflect on their own thoughts and decisions, and to be open to learning from their mistakes.

- Encourage autonomy:

  Encourage people to make their own decisions and take ownership of their actions.

Call to Action

As an AI hero, you have a responsibility to ensure the algorithms you create or use are developed and applied ethically. This call to action provides prompts to guide you in putting algorithmic auditing into practice.

Make reviews a regular habit. Convene diverse teams to inspect data and code. Listen to marginalized voices. Have the courage to make difficult changes and to address biases uncovered. And nurture a culture of transparency, accountability, and inclusion. This is hard work, but necessary. By auditing algorithms and upholding ethical AI principles, you can build systems that empower everyone. Use the prompts below to reflect on your role and develop plans to advance algorithmic ethics.

Reflect on biases	• How might my own biases unconsciously influence algorithm design?
	• What bias training can I participate in?
Diversify data	• What steps can I take to ensure training data represents diverse populations?
	• Whose voices may be missing?
Test for bias	• What user groups can I engage to thoroughly test for biases or unfair outcomes?
Monitor algorithms	• How can I implement ongoing bias-monitoring processes to identify issues early?
Assemble review team	• Who should be on the ethical oversight team to regularly review algorithms?
	• How can I seek broad input?
Evaluate transparency	• How can I analyze algorithm transparency and make processes inspectable?
Assess privacy practices	• What can I do to evaluate and enhance privacy protections around data handling?
Analyze dependence	• What is the appropriate level of dependence on algorithms for critical decisions?
	• How can I ensure human oversight?
Engage stakeholders	• How can I engage my community in discussions of algorithm ethics and get their input?
Develop action plans	• What goals and time lines can I set to improve algorithm ethics in my context?
Commit to regular reviews	• How can I make algorithm ethics reviews a regular practice and stay current on responsible AI?

Reflection: Create a Learning Hero Journal Entry

1. Write a brief journal entry summarizing your key takeaways from this chapter.

_____

_____

_____

_____

_____

_____

_____

_____

_____

_____

2. Reflect on any new perspectives or insights you gained about algorithms and AI.

Chapter 6

The Ultimate Quest: Building an AI-Driven World for All

Governments were established as beacons of guidance, even though they may falter at times. They represent our shared desires and needs. The ultimate quest of our heroic endeavors lies in actively constructing a powerful AI-driven world that benefits every single individual, without exception. Imagine a world in which the need for this AI-hero quest fades away, one in which building an AI-driven society becomes the new norm. Just as human rights transcend language barriers, the AI hero's ultimate goal is to embody the truth of our world. When we speak of government, we speak of ourselves—the people. We are the collective force that ensures regulatory fairness and ethical practices across the globe.

This ultimate quest begins with eliminating technology poverty and bridging the digital divide, ensuring that access to technology is not a luxury but a fundamental necessity for all who seek it. As we well know, technology is the gateway for active participation and contribution in today's world. As a global community, united under the banner of "we the people," we must recognize the paramount importance of regulating AI to uphold ethical, fair, and responsible use for all humans, not just the privileged.

The great news is that governments around the world are implementing laws and regulations specifically tailored to algorithms and data, aiming to prevent discrimination, safeguard privacy, and foster innovation. Agencies and task forces are being established to oversee compliance and enforcement. Some governments are actively working on policies that mandate transparency and accountability in algorithmic decision-making. It is incumbent upon leaders to grasp their role in shaping AI policies, safeguarding human rights, and protecting the interests of their constituents.

The future demands continuous evaluation and adaptation of legislation to keep pace with the ever-evolving digital landscape. We must advocate for unbiased, ethical, and beneficial AI development. As leaders, we are responsible for driving systemic change and guiding AI towards a future that aligns with our shared values and aspirations. In this ultimate quest, we have the power to shape an AI-driven world in which equality reigns, every voice is heard, and the transformative potential of technology is harnessed for the betterment of humanity.

In recent developments, government bodies are increasingly recognizing the need for AI regulation. OpenAI CEO, Sam Altman, testified before Congress, stressing the importance of setting limits on powerful AI systems to mitigate risks and protect the world from potential harm. Policymakers worldwide are engaged in fervent debates on the advantages and disadvantages of regulating AI, and the critical question lies not in the how or when, but in determining the leaders who will shape the trajectory of AI's impact on the global economy. As incumbent businesses grapple with the disruptive potential of AI, it is essential to establish mechanisms, such as nongovernmental regulators, audits, and certification processes, to incentivize ethical and trusted AI products and services.

While government regulation plays a role, self-regulatory bodies and industry standards have proven effective in the past. Collaborative efforts among businesses and academics can develop systems that identify ethical AI solutions.

Government policies are important in technology development for a number of reasons. One reason is that they can help to ensure that technology is used in a way that is ethical, fair, and responsible. For example, government policies on the use of algorithms and data may aim to prevent discrimination, protect privacy, and ensure that individuals are treated fairly.

Government policies can promote innovation and competition in the technology sector by establishing a level playing field and setting out clear rules for companies to follow. This can encourage companies to invest in research and development, and can help to prevent the misuse of technology by dominant players.

In addition, government policies can help to address potential negative impacts of technology on society, such as job displacement or the concentration of power in the hands of a few large tech companies. By setting out clear rules and guidelines for the use of technology, governments can help to mitigate these risks and ensure that the benefits of technology are widely distributed.

Finally, government policies can help to build trust in technology and encourage its adoption by the public. By establishing clear rules and oversight mechanisms, governments can help to reassure the public that technology is being used responsibly and ethically.

Governments around the world have taken a variety of approaches to regulating algorithms and the use of data. In some cases, governments have implemented specific laws and regulations to address the use of algorithms in certain sectors or for certain purposes. For example, laws governing the use of algorithms in the criminal justice system or in the hiring process may aim to prevent discrimination and ensure fairness.

In other cases, governments have implemented more general laws and regulations that apply to the use of algorithms and data more broadly. For example, data-protection laws may set out rules for the collection, use, and storage of personal data, while consumer-protection laws may regulate the use of algorithms in areas such as advertising and marketing.

Some governments have also established agencies or other bodies to oversee the use of algorithms and data, and to enforce relevant laws and regulations. For example, in the European Union, the General Data Protection Regulation (GDPR) is enforced by national data-protection authorities, while in the United States, the Federal Trade Commission (FTC) has authority to regulate the use of algorithms and data in the private sector.

Overall, government regulators play a key role in algorithmic accountability because philosophers of enlightenment in any nation, like the authors of the Declaration of Independence and the Constitution in the United States, thought of human rights as being "life, liberty, and the pursuit of happiness." They limited the role of government to the protection of these rights. Thus, the function of government was not to create or provide rights, but rather to protect us from those who would violate them.

In the United States, the White House, under President Joe Biden's leadership, issued a blueprint for an artificial intelligence bill of rights, which depicts five pillars and principles to ensure safe and effective use of technology and data (https://www.whitehouse.gov/ostp/ai-bill-of-rights/). In addition, the City of New York became the first city to assemble a task force to study, through the lens of equity, fairness, and accountability, possible biases in algorithmic decision systems.

A recent *Harvard Business Review* article pointed out that some forward-thinking regulators in the European Union's General Data Protection Regulation (GDPR) are working on policies to require that organizations explain their algorithmic decisions.

Governments have the duty and responsibility to prevent algorithmic bias and incentivize ethical usage and civil rights. Legislation must be kept up-to-date to reflect developments in our digital world. As a leader, I challenge you to reflect on your position as a citizen of this planet and to realize that you *are* the people. We are all representatives of our governments. We the people should embrace the development of leadership and policies, whether internal or external, that celebrate our differences; we should conduct full-scale audits on the emerging innovative opportunities for the benefit of all.

It is also important for elected and appointed leaders to recognize that they are accountable to the people they serve, and that they are ultimately responsible for making decisions that reflect the values and interests of their constituents. This means that, as leaders, we should be mindful of our role as representatives of the people, and we should strive to make decisions that are in the best interests of the community as a whole.

The time is now to dig deeper and commit to driving systemic change in AI development. AI has the potential to transform many aspects of society, including how we work, learn, and interact with one another. As such, it is important that AI development be guided by a set of principles and values that ensure that the technology is used in a way that is ethical, fair, and responsible. We need to call for policies and procedures that move us beyond a best-practices approach and into holistic approaches that embrace taking a more forward-looking and innovative approach to problem-solving; we must be willing to consider and experiment with new approaches that have not yet been fully tested or proven. This is especially important in our fast-moving field of technology, where new developments and innovations are constantly emerging.

Call to Action

We just explored the crucial role of governments in shaping the future of AI for the benefit of all individuals. Now, let's turn these ideas into call-to-action steps:

1. **Educate and Advocate**

   • Learn about AI: Start by educating yourself about artificial intelligence, its potential, and its challenges. Understand how it impacts society, ethics, and human rights.

- Advocate for Ethical AI: Advocate for the responsible use of AI in your community and beyond. Engage in conversations, write articles, and speak at local events to raise awareness about the importance of ethical AI development.

2. **Engage with Government**

   - Contact Local Representatives: Reach out to your local and national government representatives. Express your concerns and priorities regarding AI regulation and ethics.

   - Support AI Regulation: Encourage policymakers to support and implement regulations that promote fairness, transparency, and accountability in AI systems.

3. **Promote Access to Technology**

   - Support Digital Inclusion: Volunteer or donate to organizations working to bridge the digital divide. Advocate for policies that ensure access to technology for all, regardless of socioeconomic status.

4. **Stay Informed**

- Stay Updated: Continuously stay informed about the latest developments in AI, including regulatory changes, technological advancements, and ethical concerns. Join AI-related newsletters and forums to keep up-to-date.

5. **Collaborate and Innovate**

- Join Industry Initiatives: If you work in the tech industry, consider joining or supporting industry initiatives that promote ethical AI development and standards.

- Collaborate with Academics: Foster collaborations between businesses and academic institutions to develop ethical AI solutions.

6. **Encourage Transparency**

- Advocate for Transparency: Support efforts to require organizations to explain their algorithmic decisions, particularly in areas that impact people's lives, such as finance, health care, and hiring.

7. **Hold Leaders Accountable**

- Engage with Elected Leaders: Hold elected leaders accountable by regularly communicating your expectations for ethical AI development. Attend town hall meetings and engage in discussions about AI policies.

8. **Conduct Audits**

   - Promote Audits: Encourage organizations to conduct regular audits of their AI systems to identify and rectify biases and ethical concerns.

9. **Support International Cooperation**

   - Advocate for Global Collaboration: Encourage international cooperation on AI regulations and standards to ensure a consistent and ethical approach worldwide.

10. **Embrace Next Practices**

   - Innovate Ethically: Encourage the adoption of next practices in AI development. Be open to exploring new, innovative approaches that prioritize ethics and fairness.

Reflection: Create a Learning Hero Journal Entry

1. Write a brief journal entry summarizing your key takeaways from this chapter.

_____

_____

_____

_____

_____

_____

_____

_____

_____

_____

_____

2. Reflect on any new perspectives or insights you gained about algorithms and AI.

Conclusion

Embrace Your Heroic Destiny

As our journey nears its end, we reflect on the knowledge and insights gained throughout this adventure. We emphasize the power you hold as an AI hero and the difference you can make. Remember, the quest for inclusivity and ethical AI is ongoing, and your heroism will continue to shape the future. Go forth, mighty hero, and let your AI-powered endeavors pave the way to a world in which everyone can thrive.

The first four industrial revolutions brought about major technological and societal changes, including the steam engine, electricity, and the rise of the internet. We are living at the dawn of a new era—the fifth industrial revolution, also known as 5.0 or the smart industry—one that is expected to have a similarly transformative effect, and is already having a significant impact on many industries.

One key aspect of this revolution is the integration of human and machine collaboration, as advanced technologies such as AI and robotics are increasingly being used to augment and enhance human capabilities. This is leading to the development of new roles and job opportunities, as well as the need for team members and collaborators to acquire new skills and adapt to new ways of working.

Our lives are consumed by automation and data exchange in manufacturing technologies, including the internet of things (IoT) and artificial intelligence. Overall, the fifth industrial revolution, thanks to the development of algorithm and artificial intelligence, has brought about significant changes to the way we live and work, and will likely continue to impact our lives in major ways.

Algorithms have taken our human existence to another dimension. They have permeated every single aspect of our lives, from health care to education, our homes, transportation, and even laws. They have changed the way we work, the way we live, and the way we relate to one another and the world. And in some cases, historians are calling this the era when humans became hackable.

However objective we may intend our technology to be, it is ultimately influenced by the people who build it and the data that feeds it. The challenge is figuring out the steps we need to take to harness and realize the benefits in the long term and also realize the larger responsibility of technological innovation, where we hold each other accountable and measure societal impact alongside financial performance. Algorithms are amazing, and we need them. But let's not forget that people create algorithms, and their biases can inadvertently influence the outcomes.

As a leader in your community, you are the key to our future success. Review and analysis of developments should be routine to ensure that AI systems conform to the conventions established at the organizational, societal, and governmental level. In most cases, your role may not be to simply dictate what is fair. Rather, you will guide the assessment and provide access to opportunities to correct bias within the parameters set and governed by the law of the organization's understanding of what responsible AI is. This is the step necessary to build AI trust with its collaborators, avert risks, and contribute value to society.

As former President of the United States, Barack Obama said, "Don't just play games on your phones. Program the game." And if I may add as a leader, "Get in the game to support the programming and power up the algorithm."

We all have a social responsibility to hold technology developers and organizations accountable, and by learning how algorithms work, leaders can gain the skills to play a role in the development of systems and applications that are benefits to society. I encourage you to not just consume technology, but to actively engage with it and learn how it works, in order to be involved in creating it ethically.

Leaders should support and encourage the development of these skills, perhaps by providing resources or opportunities for individuals to learn more about how to power up the algorithm. By doing so, leaders can help to ensure that the next generation of collaborators have the knowledge and expertise to drive technological innovation and progress.

Algorithms and AI are the next steps in an evolution, and we are just at the beginning. They are still in the early stages of development and have the potential to evolve and advance significantly in the future. As we have learned, there is still a lot of room for improvement and further development. It is likely that as AI technology continues to evolve, it will have an increasingly significant impact on many more aspects of our lives.

For better or worse, I think it's clear that AI is here to stay. The choice is not when, but how do we join the evolution to make it better for all. The use of algorithms and AI is not a matter of choice, but rather a reality that we must confront. The question is not whether we will use these technologies, but rather how we will use them in a way that benefits everyone. We have a responsibility to actively engage with the evolution of algorithms and AI, and to shape their development in a way that is positive and beneficial for society as a whole.

Let's all be part of shaping the changes and the future we wish to see. Let's all work together to power up the algorithm.

Reflection: Create a Learning Hero Journal Entry

1. Write a brief journal entry summarizing your key takeaways from this chapter.

_____

_____

_____

_____

_____

_____

_____

_____

_____

_____

_____

2. Reflect on any new perspectives or insights you gained about algorithms and AI.

Appendix

Tools, Worksheets, and Resources for an AI Hero

In our unwavering commitment to your hero's journey, we present a treasury of invaluable tools, interactive worksheets, and a wealth of additional resources. These carefully curated assets will serve as your guiding companions, enabling you to seamlessly translate the concepts explored in each chapter into impactful actions. Equipped with these transformative resources, you are poised to lead with purpose and navigate the ever-evolving realm of AI.

The time has come to embrace your extraordinary AI heroism. Together, we shall embark on a thrilling adventure, weaving a tapestry of possibilities that harnesses the true potential of technology. Let us forge a future in which innovation becomes an empowering force, uplifting humanity and celebrating the glorious tapestry of our diverse world. The stage is set, and the world eagerly anticipates the heroic transformation you are destined to unveil.

For all tools, worksheets, and resources, please go to www.powerup.org/ai.

About PowerUp.Org

PowerUp.org is a nonprofit organization dedicated to bridging the digital divide in our community! Our programs are dedicated each year to providing early-childhood-to-career programs aimed at inspiring and educating thousands of students from underrepresented and underserved communities to try computer science and coding.

PowerUp.org is organized exclusively for charitable purposes. We are an IRS-approved, tax-exempt organization under section 501(c)(3) of the Internal Revenue Code. PowerUp.org stands for "Providing Opportunities Where Everyone Rises Up." The purpose of this organization is:

1. To support and conduct nonpartisan research and education, ensuring that all children—especially those at risk of being left behind—have the resources and the opportunities they need to grow up healthy and lead productive lives.

2. To provide informational education-based activities to increase public awareness of creating technology-rich learning environments that inspire children to embrace

their individual potential and engage with the fast-moving world around them.

3. To develop and discover ways to connect more children with the tools that will inspire them to develop into engaged global citizens.

Our Mission

Our mission is to ensure that all children—especially those at risk of being left behind—have the resources and the opportunities they need to grow up healthy and lead productive lives. The motivation behind PowerUp.org is to raise the profile of technology education to young children. It has three specific aims:

1. To showcase and promote computer science education as an important and exciting tool for community success.

2. To provide the community with a new approach to motivating and enthusing students to learn computer science.

3. To engage students in science, technology, engineering, and math (STEM), as these are the most important factors in creating tomorrow's workforce.

Our Services and Programs

PowerUp.org programs are centered around computer coding. We provide opportunities in which everyone rises up as a coder. We are particularly focused on making sure more students from underrepresented and underserved communities give computer science a try because research shows they are far more likely to pursue computer science later in their academic career if they have had some early exposure to it.

PowerUp.org will accomplish its goals by conducting four major activities:

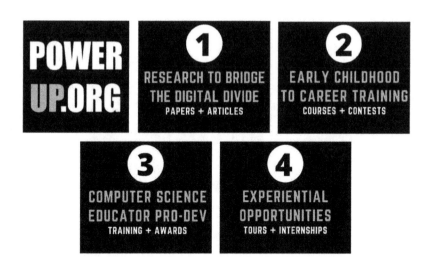